CAPE COD

A VOLUME IN THE SERIES
Environmental History of the Northeast
EDITED BY
Anthony N. Penna
Richard W. Judd

CAPE COD

An Environmental History
of a Fragile Ecosystem

John T. Cumbler

UNIVERSITY OF MASSACHUSETTS PRESS
Amherst and Boston

Copyright © 2014 by University of Massachusetts Press
All rights reserved
Printed in the United States of America

ISBN 978-1-62534-109-9 (paper); 108-2 (hardcover)

Designed by Sally Nichols
Set in Adobe Garamond Pro and AiKochAntiqua
Printed and bound by Sheridan Books, Inc.

Library of Congress Cataloging-in-Publication Data

Cumbler, John T.
  Cape Cod : an environmental history of a fragile ecosystem / John T. Cumbler.
      pages cm. — (Environmental history of the Northeast)
  Includes bibliographical references and index.
  ISBN 978-1-62534-109-9 (pbk. : alk. paper) — ISBN 978-1-62534-108-2 (hardcover : alk. paper) 1. Human ecology—Massachusetts—Cape Cod—History. 2. Nature—Effect of human beings on—Massachusetts—Cape Cod. 3. Cape Cod (Mass.)—Environmental conditions. I. Title.
 GF504.M4C86 2014
 304.209744'92—dc23
                                    2014019830

British Library Cataloguing-in-Publication Data
A catalogue record for this book is available from the British Library.

To Judith Cumbler

CONTENTS

Acknowledgments   ix

Introduction   1

PART I
The Ice, the Crow, and the Plague
Before European Exploration

1. From Continental Drift to Nomadic Land Use   13
2. Fire, Fishing, and Farming of Native Peoples   24

PART II
The Era of Local Resource Production and Extraction
Settlement to the Start of the Twentieth Century

3. On the Way to an Amphibious Society   31
4. Mining the Bounty of Nature   51
5. The Decline of the Established Economy   80

PART III
Dependence on Distant Resources, and Revenue from Recreation
Early Twentieth Century to the Present

6. Trains, Cars, Cottages, and Restaurants   113
7. The Golden Age of Tourism   149
8. Problems in Paradise   173

Conclusion   198

Further Reading   211
Notes   215
Index   265

ACKNOWLEDGMENTS

Most authors depend on other people to help them get their work to a place where they want it to be. That is certainly the case for me. An acknowledgment can never fully capture all the help I received in moving this project toward publication. My book is all the better because of that help.

As is the case for all of my work, Sam Bass Warner has been an inspiration, a mentor, a friend, and a very helpful critic from the beginning of this project to its final form. Tony Penna and Dick Judd are the kind of series editors for which any author would long. Brian Halley has patiently helped move the manuscript along. Pamela Mack and her mother shared with me their Underwood family history (some of which appears in text and pictures included in the book). Jim O'Connell made vital suggestions about how to improve the last two chapters. Chad Montrie, Donald Worster, and Eileen Palmer read sections of the manuscript and added helpful suggestions for improvement. Readers will appreciate that Stu Galley, Martha Gordon, and Jim O'Brien read over most of the manuscript with an eye to improving the flow of the narrative. Steve Miller helped with a difficult chapter.

Although it would be hard to find a more wonderful place to do research than Cape Cod, it was made even more pleasant by the aid of a number of librarians. The librarians in the Nickerson Room of the Cape Cod Community College library were helpful and cheerful. Besides the collection housed at that library, the other major collection of Cape-wide material resides at the Cape Cod National Seashore, Park Headquarters. The park has a rich collection of research documents and scientific reports. The helpfulness of the park's staff, particularly the collections librarian, eased the transition from the beauty of the bike trail I took to get to the library to the work ahead in the documents room of the headquarters. Across the Cape are local libraries, each with a rich collection of material. I was able to access this material though my own local Wellfleet Library. The Wellfleet librarians, especially Elaine McIlroy and Naomi Robbins, were helpful, cheerful, and encouraging of this project. My understanding of the local environment was greatly aided by John Portnoy. Discussions with Brian Donahue and Matthew McKenzie helped clear my head of misconceptions about both farming and fishing.

This book is also a product of summers on the Cape with my children, Kazia and Ethan, my children-in-law, John and Jane, and now my grandchildren, Rae Low and Emmett Cumbler. The book is dedicated to Judith Cumbler, who has always been part of my Cape Cod experience.

CAPE COD

INTRODUCTION

Few of us today who visit Cape Cod can remember the Cape that is celebrated in old books and songs and in pictures and artifacts on the walls of motels, inns, and restaurants. Yet it was only little more than a hundred years ago when some of the central participants in that older Cape Cod world began to realize that the era of fishermen, sailors, and farmers was ending, and a new era was being born. One of those people was Lorenzo Dow Baker, the famous banana captain from Wellfleet, a town on what locals called the "Lower Cape" because it lay downwind from Cape Cod's main body.[1]

In 1906 Baker went before the Wellfleet town meeting to argue that the town should spend tens of thousands of dollars diking the tidal inlets to the town's saltwater marshes in order to lower the mosquito population and encourage tourism. This was a big request, but Baker was an important town leader whose voice carried weight.

Lorenzo Dow Baker was the son of a Wellfleet fishing-schooner captain and a fourth-generation fishing sailor. At age eleven he had shipped out on his father's vessel as a cabin boy, and by his twenties he owned his own seventy-ton schooner. In 1871 his vessel, the *Telegraph of Wellfleet*, limped into a Jamaica harbor after battling a hurricane en route to the Cape from Venezuela. Looking for a profitable cargo to make up for the costs of repairs, Baker took on a load of bananas. After selling the bananas in Jersey City, Baker returned to Jamaica for more. These he sold to a fruit merchant in Boston who would soon become his partner in the Boston Fruit Company, which in 1888 emerged as the dominant force in the newly formed United Fruit Company. Baker had not favored the merger (or the larger plantation model that would preclude buying bananas from small patch holders), but when he withdrew from the company he took with him stock holdings worth millions.[2]

Baker was born during the golden years of the Cape, when the masts of more than a hundred fishing schooners filled Wellfleet Harbor like a forest of leafless trees. Whaling ships brought home whale oil from the North Atlantic. Acres of racks (known as flakes) with drying fish stretched out behind the shore while windmills pumped saltwater into dozens of evaporating vats. Coopers, shipwrights, boat repairers, sail makers, caulkers, and merchants crowded about the busy wharves. Farmers brought in milk, eggs, and vegetables to feed the busy harbor town. But the Cape to which Lorenzo Dow Baker returned after his trips between Jamaica and Boston was not the prosperous one in which he grew up. Of the hundred fishing schooners that sailed out of Wellfleet in 1850 only twenty remained in 1890.[3] Wharves from Truro to Falmouth rotted, and Cape shipyards stood empty. Scavengers dismantled the old saltworks for the wood, and farmers found little market for their produce.

Baker knew the Cape and he knew fishing. He had fished most years before his run to Venezuela, and the winter he returned from his successful banana run to Boston, he took the *Telegraph* out for another fishing season. The rewards for his winter of fishing of 1871–72 were so small, however, that he forfeited the owner's share of the profits in order to give his sailors a share, even if woefully small, for their labors.[4] His childhood friends struggled even harder to survive as fishermen or farmers. Their children left the Cape for opportunity elsewhere, and homes fell into disrepair. The forests that had supplied the wood for boats, saltworks, barrels, and fuel had been stripped from the land, and sand now blew across abandoned fields and silted-in harbors.[5] As a result, Lorenzo Dow Baker formed a different vision for the Cape's future.

The Jamaica banana trade that Baker initiated required a massive infrastructure. At Port Antonio, where Baker gathered his bananas, he oversaw the building of wharves, warehouses, a hospital, schools, boarding houses, and a large hotel for company officials to stay. Looking for an opportunity to expand employment for his workers during the off-season and keep the hotel full over the winter, Baker advertised the hotel as a tourist retreat.

Soon thereafter he brought this experience to his hometown of Wellfleet. Here he bought a fishing wharf on a beach by the town harbor in the mid-1880s, and in 1902, transformed it into the massive Chequessett Inn which stretched out over the water. Baker brought up workers from his Jamaica hotel and combined them with college students and locals to staff the inn during the

INTRODUCTION

By the early decades of the twentieth century the Cape had become a resort for middle- and upper-middle-class Americans. Photo courtesy of the Library of Congress.

summer season. The Chequessett Inn was a success. It was also part of Baker's vision for a different Cape Cod—a place for tourists and vacationers to spend money enjoying sand, surf, and sea in the midst of old Cape character and the local heritage of farming and fishing.

Baker realized the attraction of the Cape depended upon its natural beauty, but visitors would need to be comfortable and safe if they were to enjoy it. In Lorenzo Dow Baker's new vision the bay and harbor, which had sheltered fishing and trading boats for two hundred years, would provide a safe place for sport fishermen and pleasure sailors. Wellfleet's shallow creeks and river, surrounded by acres of marsh grass, fed the harbor. These streams sustained rich fish populations, attracted spring herring runs, and provided the fresh water that nourished the town's famous oyster beds. The marsh grass in turn provided fodder for the town's livestock. But the tidal waters flooded and ebbed so that the marshes flourished with life, and unfortunately, insect life. Even the tough naturalist Henry David Thoreau had complained of the mosquitoes on the Cape.[6] Lorenzo Dow Baker understood that his new vision of Cape Cod could prevail only when nature—more specifically, the insect population, especially "mosquitoes that now infest the town"—was controlled. It was the insect problem that brought Baker to the Wellfleet town meeting.[7]

Baker argued at town meetings from 1906 to his death in 1908 in favor of diking the river and draining the marshes. He argued that once the mosquitoes had been eliminated, more land would be available for vacation homes that would appeal to off-Cape residents. In response, the town began building dikes but balked at the largest one, the one intended to close off the tidal Herring River, which flowed in and out of Wellfleet Bay. It would cost the town twenty thousand dollars and would mean a dramatic increase in town expenses, when the total budget of that year was only fifteen thousand dollars. It would also reduce the modest revenue the town derived by auctioning off the right to fish river herring upstream from the proposed dike.[8] Baker and his allies pressed the advantages of the new tourist economy. Baker brought in experts who argued that the money spent on the dike would be made up in the increased value of the property, as the land "would at once [be] in demand for sites for summer residents, were the mosquitoes reduced to a normal quantity and every homestead would have a greater selling value."[9]

Not all the town's residents were convinced. Above the mouth of the Herring River where the dike would be constructed, rested numerous oyster beds that were worked by local oystermen. Here, too, shore fishermen had their fishing shacks to keep their boats and equipment. For almost two centuries the spring herring run up river had supplied ample bait and food for the town's fishermen and residents.[10] More than two hundred thousand herring were harvested from the river between 1888 and 1890, making it the second most productive herring harvest area in the state. Residents argued that the herring runs up the river were too important to the town and its history to be destroyed, and that the livelihood of those upstream should be considered.

In 1903 Baker had thrown a massive clambake on Billingsgate Island, where the whole town filled up on free lobsters, oysters, clams, corn, and Jamaican fruit. Such generosity established Baker's status as local icon. Despite his popularity in town, resistance to the Herring River dike held until 1908, the year of his death. As Wellfleet mourned its famous sea captain and wealthiest resident, Baker's supporters argued that the final dike on the Herring River would be the best tribute to him. That appeal was too strong for the majority of the dike's opponents to resist, and the town meeting voted its approval.

Twelve of those who opposed Baker's vision of their town as a tourist destination brought suit against the dike, arguing that the destruction of the herring runs violated the common-law rights of the citizens to fish. When the town promised to open the dike's gates to allow the migration of herring, the case was dismissed, and the dike was built. Yet those who fought the dike were right to be concerned. Although the gates were opened, so few fish made it to their spawning ponds in 1910 that the town could get only one bid (of just seventy dollars) for the fishing rights. The next year no one bid, because no fish made it past the dike. All the fishermen, oystermen, and clammers above the dike had to look elsewhere for employment.[11]

Turn-of-the-nineteenth-century Wellfleet can be seen as a microcosm of the entire Cape. When Henry David Thoreau visited the Cape on four occasions in the mid-1800s, he called it a "weather-beaten garment" with "many holes and rents."[12] Yet when he trekked across Cape terrain on those four trips, Cape Cod was just beginning to feel the consequences of two centuries of heavy environmental use. Although the Pilgrims may have been exaggerating when they claimed to have found thick forests of pine, oak, beech, hickory, sassafras, maple, birch, ash, hickory, cedar, and holly, they clearly did confront a forested Cape. By the time Thoreau came to the Outer Cape a mostly treeless vista confronted him. The lack of trees caused the soil to erode. Blowing sand and poor soil plagued Cape farmers who also struggled with low prices and constrained markets.[13] The sea's seemingly unending fecundity grew more and more uncertain. The salt marshes that had provided crucial fodder for livestock for almost three hundred years contributed little of importance to a world of steam railroads and milk imported from off-Cape dairies. Lorenzo Dow Baker and others believed the key to the recovery of the Cape's economy and its environment lay in the new economy of tourism. Farms would become inns, while shoreline saltworks, fish flakes, and fish houses would give way to vacation cottages. Pastures would become croquet fields and tennis courts. Fishing piers would become docks for pleasure boats and sport-fishing vessels. Local farmers would supply food for vacationing visitors.

The economic recovery that Baker envisioned did begin soon after his death. In the 1920s the Cape's population began to increase, and by the 1950s land values rose dramatically. Over time a new forest began to grow and beach grass held down blowing sand.[14] But the new Cape economy was not without social and environmental problems. Like those who lost their livelihoods when the Herring River was diked, some on the Cape failed to prosper with the new tourist economy. And, although the land appeared to recover with the return of forests, environmental problems of water pollution, congestion, and different forms of soil erosion emerged with the tourist economy.

During the last four thousand years Cape Cod has experienced three distinct regimes of resource utilization.[15] The first regime, which dominated the Cape before European exploration, was a Native American regime. It began with periodic visits by mobile hunting people in the Late Archaic era when the Cape was just beginning to take its modern form. By the time of European contact four hundred years ago, fire, fishing, shellfish harvesting, and horticulture came to typify Native American resource use on the Cape.[16]

The second regime extended from the middle of the seventeenth century to the beginning of the twentieth century. It was a regime of resource extraction and production: farming, fishing, boatbuilding, and salt making. This second regime began as an amphibious era of sail and water but gradually accommodated such new phenomena as steel, steam, petroleum, and electricity. Cape Cod is presently in the midst of the third regime, a regime of distant resources and a local economy of recreation and tourism.

Each of these regimes treated the Cape's fragile environment in different ways. The effects of the three regimes upon the environment of Cape Cod are the central concern of this book. Of course large forces of weather and climate have shaped and continue to shape the Cape ecology.[17] This book will also look at how human decisions and actions interacted with these larger nonhuman forces. Specifically, the book proposes that during the second regime of local extraction and production the Cape's environment slid into a crisis whose consequences, a stagnant economy and a stressed ecosystem, endured for seventy-five years. The economic crisis was ultimately resolved by the emergence of the new regime dependent on distant resources and the revenue from recreation and tourism. Over the next fifty years the Cape's economy flourished, but this also took its toll on the region's ecosystems. Problems of this new post-productive industrial age continue to haunt the Cape, socially, economically, and environmentally; much of the Cape's modern history involves attempts to come to terms with these problems.

Change typifies each of these three periods. During the first regime the land changed significantly, partly though human agency but even more so because of larger geological forces. The people who came to use the resources of the Cape also changed; their changing in turn affected the environment within which they lived, though little evidence remains for us to understand whether people who occupied the Cape before European contact pushed their environment toward any kind of crisis. For the next two regimes we have significantly more evidence, and that evidence suggests that resource use between the first white settlement and the early nineteenth century stressed the Cape's environment but did not overwhelm it. Forests were cut and blowing sand and erosion plagued farmers and silted-in harbors, but Cape Codders were adaptable. They substituted peat for cordwood, for example, and they abandoned traditional crops for turnips, asparagus, and other plants that thrived in sandy soil. They converted marshes to cranberry bogs and utilized the blowing sand to spur berry production. They increased their harvest from the sea, using new and more aggressive techniques. But by the middle of the nineteenth century the environmental stresses multiplied and the options for adaptation shrank. Old systems of making a living were abandoned and communities lost people and opportunity.[18]

The environmental and economic strains that led to the Cape's decline in the second half of the nineteenth century did not spell continued decline but rather, transition and adaption. The shift to a recreational-tourist economy allowed for the reforesting of the Cape's landscape. But just as past land uses stressed the Cape's fragile environment, so too did the present regime of tourism and recreation have its environmental costs. Just as people in the past tried to change and adapt, people today work to mitigate the negative consequences of their present land use. The environmental history of the Cape is a story, not of relentless declension, but of change.[19]

While viewing the long span of the Cape's human history through the lens of three regimes of resource utilization, it is important to remember that the process of change has been complex and overlapping. At each juncture, older means of resource utilization continued into the new era just as new practices had their roots in previous periods. Also, central to this book is the understanding that in each regime there are those who have greater control over resources and the ability to decide how they will be used, and there are many who have less control. During periods of regime transformation there are winners and losers. Large environmental and economic forces drive the history of Cape Cod, but local human agency directs the outcomes.[20]

Although there was a certain commonality of experience across the Cape, the class, gender, occupation, and social position of the residents brought differing experiences. The burden, for example, of the shift from the extractive productive economy to the tourist vacation economy was weighted unevenly across different sectors of the Cape. As family income from fishing or farming declined, initially it often fell to the female members of the household to make up the slack by entering into service as cooks, maids, or cleaning staff in the new motels, hotels, guest homes, and estates while the men continued to try to wrest a living from traditional activity. Small farmers on more marginal land were the first to feel the pressure of an exhausted environment. Some boat builders successfully switched to pleasure crafts, but others had to take their skills elsewhere. The new diesel trawlers and new markets provided a comfortable living for Provincetown fishermen who could afford the new boats or find berths on them, but the crews of the new trawlers made up a mere a faction of those on the old fishing schooners, and the fishermen left ashore needed to look elsewhere for work.

Residents also experienced changes as events within their own communities. Because of the significance of town life, this story will weave between the Cape as a whole and the individual towns. The Cape is one county, Barnstable, but the real center of life, politics, and power until the late twentieth century rested at the town and state (or, earlier, colonial-government) levels.

By 1849, when Henry David Thoreau made his first trip to Cape Cod with his friend and frequent walking companion William Ellery Channing, a well-known Transcendentalist poet, Thoreau had written only one book, *A Week on the Concord and Merrimack Rivers,* which he completed while at Walden Pond. It included ponderous digressions on a number of topics and was not a success. Perhaps chastened by that, Thoreau wrote about Cape Cod with wry self-reflection and dry humor. *Putnam's Magazine* published the four pieces based on Thoreau's time on the Cape. But Thoreau's more ironic style led to a falling out with the editors. Eventually Thoreau's writings on the Cape were posthumously published, through the work of Channing and Thoreau's sister Sophia, as *Cape Cod.*

The Cape Cod of which Thoreau wrote was a sandy, watery place where traveling by foot or boat was easier than by coach; the train had only just arrived at Sandwich when Thoreau and Channing made their first 1849 trip to the Cape. The Cape at which Thoreau and Channing arrived was an overworked, changing, and, except for pockets along the west–east line of the Cape, mostly treeless environment.

The basic geography of the Cape is relatively new, having been set down some fourteen to eighteen thousand years ago. The physical Cape that Thoreau described mostly remains, although today much of the Outer Cape path Thoreau walked in 1849 is a hundred yards out to sea. The Cape is no longer mostly treeless, but it is still sandy and watery. The towns Thoreau walked through are with us today and are still the center of life on the Cape.

INTRODUCTION

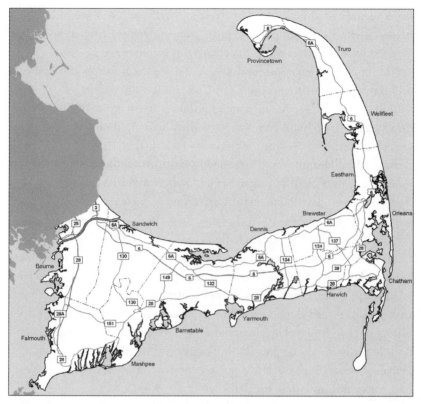

Map of Cape Cod. Courtesy of the Cape Cod Chamber of Commerce.

Although glacier deposits created the Cape, its topography does vary and that affects the history of the different towns. The results of differences in the way heavier and lighter glacier deposits eroded and moved, and the differences in the nature of erosion and its effects on deposits of sand, clay, and soil fostered different plants and distinct ecosystems. The sand spit off of Barnstable created an environment conducive to the creation of a massive salt marsh. Fresh water flowing into Wellfleet Harbor from the Herring River encouraged, not only a large salt marsh, but also a fecund environment for shellfish. Migrating sands and erosion of land from glacier holes created better harbors and easier access to the fishing and whaling grounds for Provincetown, Wellfleet, Chatham, Barnstable, Dennis, Yarmouth, and Falmouth than for other Cape towns. Eastham, Brewster, Dennis, Yarmouth, and Barnstable had richer soil than many of their neighbors, but Barnstable and Yarmouth had more rocks and boulders with which farmers struggled. These differences affected how Cape Codders lived their lives while they shared this common spit of sand.

This book is an environmental, social, and economic history of Cape Cod told through the experiences of Cape Codders and the visitors who came to these shores. The Cape has held an important place in our national imagination for almost two hundred years. Naturalists, activists, statesman, social critics, poets, and novelists—such as Timothy Dwight, Thoreau, John Hay, Henry Beston, Mary Heaton Vorse, Edmund Wilson, Mary McCarthy, John and Katharine Dos Passos, Norman Mailer, Robert Finch, Marge Piercy, and Richard Russo—have found on the Cape an aura of mystery and fascination as well as settings for their work. As Lorenzo Dow Baker believed would happen, the Cape has indeed become the nation's vacation wonderland. But the Cape's special place in our culture should not screen us from the important historical lessons it has to tell about our environmental past and present. The world with which Cape Codders have struggled and continue to struggle is a world shared with peoples settled along many of the coasts of the globe. So too, the Cape's history of crisis from overextraction during the second half of the nineteenth century prefigures the coming crisis of overextraction of the world's resources. Perhaps if we are capable of learning something from the Cape's history we may have something important to tell the rest of the world.

I
THE ICE, THE CROW, AND THE PLAGUE
Before European Exploration

1
FROM CONTINENTAL DRIFT TO NOMADIC LAND USE

Four hundred years ago, a young child sent out to protect fields of maize was told to chase away the birds but not to hurt them. She was told not to harm the crows, in particular, because they were the givers of life. In a time before memory, her people believed, the crow brought maize and beans from the great god Cawtantowwit's fields in the Southwest. In fact, maize grew well in the soils of the peninsula that jutted out sixty miles from the mainland. The soil was sandy, but the land was thickly forested and well watered, and the organic litter on the forest floor provided a rich planting ground once the tree canopy was removed.

Compared to the billions of years in which the earth took form, the land that came to be called Cape Cod was a very recent creation. The sandy soil on which the child stood guarding the planted fields may have seemed timeless to her, but it had been deposited there only ten to fifteen thousand years before, the gift of glaciers moving on top of ancient stone.

CONTINENTAL DRIFT

The bedrock deeply buried under the sand of Cape Cod was formed in the Paleozoic era, when the early continents drifted together and thrust up mountains of plutonic rock. Some six hundred million years ago, a chain of mountains was created by the uplift of volcanic activity. This chain, known as the Avalon Belt or Zone, stretched from Newfoundland south to the Chesapeake. The drift of continents pushed this belt into the North American continent, creating a volcanic granite rise stretching from northeast to southwest across New England. As the Avalon Belt subsided, it created the Boston Basin, encompassing most of what is now metropolitan Boston, and Narragansett Basin, stretching from Providence on the Narragansett Bay to Scituate on Cape Cod Bay. Over the next few hundred million years this ancient floor of volcanic rock and granite weathered and shifted about, tilting eastward to form the miles of Atlantic shelf that encompass rich fishing grounds, such as the Grand Banks, which sustained so many thousands of fishermen for more than four hundred years.

Thirty-three million years ago, when tropical forests covered most of the land of New England, shifting landmasses in the Southern Hemisphere created the Antarctic Circumpolar Current, which cut off the pole from warmer waters. Ice began to spread out across Antarctica, reflecting sunlight away from a cooling earth. As the earth cooled, tundra slowly replaced the North American rainforests.

Two and a half million years ago North and South America collided at the Isthmus of Panama. The warm west-flowing North Equatorial waters pushed into the Gulf of Mexico and then spilled out in a northeast-flowing Gulf Stream. The Gulf Stream in turn brought warm moisture-laden air north into collision with cold arctic air, causing significant increases in the amount of rain and snow. Forests spread across the north, and snow began to pile up as more fell in the winter than could melt during the short summer. Each new snowfall compacted earlier snows into thick ice, reflecting more of the sun's energy away from the earth and furthering the earth's cooling. The accumulating snow and ice grew to glaciers more than a mile high. The pressure of the weight of the glaciers slowly pushed them south, eventually covering the northern half of North America. Shifting wind and water currents as well as the tilt of the earth affected the reach of these glaciers over the next two million years, with episodes of advance and retreat. At their farthest reach they absorbed so much water that they lowered the level of the sea four hundred feet and their weight depressed the very surface of the earth, pushing it down three hundred feet below the present elevation. These mile-high glaciers plowed across the ancient mountains of northern New England. In the process they ground up old volcanic and granite materials and pulled them into the glacier itself. Larger boulders and stones were also embedded into the glaciers, along with soil and clay plowed up from hilltops and river valleys.

The ice sheets that expanded south during the Pleistocene epoch, beginning more than a million and a half years ago, sent four major glaciers pushing across New England. After building in Canada for some fifty thousand years, the Laurentide ice sheet, the most recent of the Pleistocene glaciers, began pushing out southward roughly seventy thousand years ago, during the Wisconsin glaciation. It advanced across New England, reaching Cape Cod and Long Island twenty-one thousand years ago.

GLACIERS MELT

About fifteen to eighteen thousand years ago at the end of the Pleistocene epoch and the beginning of the Holocene epoch, the glaciers began to melt back. Over the next several thousand years, slowly and erratically, the glaciers began to break up. Initially they came to rest for roughly a thousand years along a line stretching northeast from the tip of Long Island out across Block Island, Martha's Vineyard, and Nantucket to the Nantucket Shoals. Then followed a more rapid melt back to the present Cape Cod area, where the glaciers stopped again for approximately another thousand years before their final retreat to the arctic location they held through most of the Holocene epoch.[1]

When the glaciers began to melt back they also left behind what they had plowed up as they pushed south. The melting glaciers deposited drift, also called "till." This was material washed out by running water from the bottom, sides, and front of the glacier. Large masses of till that formed at the front of the glacier were known as terminal moraines. Glacier deposits (composed of a mix of boulders, rocks, clay, sand, and gravel) washed free by streams and rivers of melting water were swept along into long drifts or eskers, filling valleys and river mouths. The Laurentide ice sheet along the southern coast of New England dumped a massive load of till into ridge mounds. These humpbacked terminal moraines mark resting points, either a glacier's furthest advance or a halting point in its retreat. These moraines formed first along Long Island, Martha's Vineyard, and Nantucket and then along Cape Cod. Long Island, Martha's Vineyard, Nantucket, and Cape Cod were not the only parts of the region to receive a heavy load of glacier till. Most of New England is covered with a layer of till and drift as meltwater left behind an outwash plain of sand, gravel, and boulders. Because the Cape's creation was part of the larger glacier construction, it can be seen as part of an archipelagic line of islands reaching from Cape Cod down to lower Long Island.

THE SHAPE AND REFASHIONING OF GLACIAL MORAINES

Three lobes of the retreating ice sheet, leaving behind massive loads of till along their moraine, formed the Cape. The Buzzards Bay lobe ran along the present coast of southeastern Massachusetts down eastern Buzzards Bay though Falmouth to Martha's Vineyard; another, the Cape Cod or Sandwich Lobe, ran eastward from the northern section of the Buzzards Bay lobe toward Dennis. A third lobe stretched north–south along a line east of the present Outer Cape. The till left behind by the Buzzards Bay and Cape Cod lobes created the terminal moraines that today provide the high points of the Upper Cape, the sites for Route 28 to Falmouth and the Mid-Cape Highway. The lobes formed a bowl, melting into a glacier lake blocked at its southern end by earlier deposits. As the glacier melted, till washed out from the sides, front, and bottom of the ice sheet. Along the moraine of the Cape Cod lobe, the till as it washed south formed a broad sandy plain stretching into Nantucket Sound. The north–south lobe, resting east of the present coast of the Cape, dumped and washed glacial till along the northern arm of today's Cape Cod.

In addition to rocks, clay, gravel, sand, and boulders, huge chunks of ice broke off from the glacier and were left behind embedded in the gravel and till. When these ice deposits finally melted they left giant deep holes. As the water table rose, the deepest holes became the five hundred freshwater kettle and saltwater ponds so loved by present-day vacationers on Cape Cod. Some of these, such as Brewster's 743-acre Long Pond, were huge, while others took up less than an acre.

The sand, gravel, and clay left behind by the glaciers lay some two hundred feet over the initial bedrock and were in constant motion for the next several thousand years. Melting glacial waters initially spread the till out from the terminal moraine. Rain, wind, waves, and currents later moved the deposits into a swath ten miles wide along the southern stretch, the Upper Cape. The erosion from the north–south lobe created a thinner band along the northern stretch, called the Lower or Outer Cape. Then, six thousand years ago, with the melting of the glaciers, the ocean waters rose. The early millennia saw the most rapid sea level rise of almost fifty feet, or fifteen meters, per thousand years.

Between two and six thousand years ago, rising sea levels slowed to just short of ten feet per thousand years. Sea levels rose another six and a half feet, finally reaching more than three to four hundred feet above the pre–glacier melt level. Rising sea levels slowly flooded over much of New England's large coastal plain. Nantucket Sound and Buzzards Bay filled with water, reaching their present level some thousand years ago.[2]

Wind, currents, and tides pushed and cut away at the sand, clay, and gravel deposits, particularly at the northern arm of the Cape. Lighter sand was pulled into tidal currents and moved about across the face of the Outer Cape. As the sand was moved about, it created new land along the northern and southern tips of the Outer Cape. With rising sea levels, eroded sands began forming barrier beaches, which protected the cliffs and filled in the area behind creating large marshes. Another two thousand years would pass before the barrier spit of Sandy Neck formed in Barnstable, allowing for the formation of Barnstable's great salt marsh. This was thirty-seven hundred years ago. Wind also lifted lighter sand and deposited it as coastal sand dunes. The increase in sea level also led to rising freshwater levels in the kettle ponds and marshes. Water flooded over plants and decaying organic material, creating the conditions for rapid peat accumulation. Vegetation, particularly trees and grasses, took root on the dunes and stabilized them; heavy storms and, on occasion, human action stripped the dunes of their cover, destabilizing them by facilitating their blowing and moving about.

THE OCEAN EXPANDS

Rising sea levels covered the hills and plateaus of the old plain. These became the various "banks" so prized by fishers. Fish flourished on these banks because nutrients in seawater tend to sink into deep water, where the absence of sunlight reduces the process of creating life. Where ocean currents collide and flow over underwater hills and plateaus, on the other hand, deepwater nutrients are brought to the surface where they get taken up by plants, using the energy from the sun. These simple plants then become the food for larger organisms. The long summer days in the Arctic and Boreal regions, the home of cooler waters of northern New England, provide sunlight for the massive production of algae and the small organisms that live off those algae. The nutrients created by these interactions flow south with the Labrador and Nova Scotia currents.

When these currents collide with the northeast-flowing Gulf Stream and flow over the shallow areas of the banks off the New England and Canadian coasts, the nutrients provide the basic building blocks of the seawater food chain. The incredibly rich fish stocks of the banks and shoals off the coast of the Canadian Maritime provinces and New England originate in this complex interaction.

CLIMATE

Two major ocean currents flow past the Cape in opposite directions. The Labrador Current flows out of the Arctic, bringing cold dense water down across the eastern front of the Cape, called the backside by locals. Flowing along the south side of the Cape is the warm water of the Gulf Stream. This huge volume of Gulf Stream water warms the southern Cape, and when the air warmed by this water comes in contact with the colder land and waters of the northern Cape, the moisture of the ensuing fogs helps maintain thick coastal vegetation even in poor soils. Waters that surround the Cape affect its temperature. In the summer, the cool waters of the Labrador Current hold down summer heat, particularly for the Outer Cape, while warmer waters of the Gulf moderate winter. Wherever one is on the Cape, temperature variations are less extreme than inland. Spring planting season comes late, but the fall harvest stretches deep into November. The influence of the boreal waters, along with the continuous flow of nutrients down from the inland hills and mountains that push close to the New England coast, feed a rich diversity of plant and animal life along the coast.

Over the summer months, while our indigenous child watches the growing maize, winds flow primarily from the southwest, bringing with them warm air. The collision of this warmer-wetter air from the southwest with the cooler air currents coming from the northwest lends itself to regular summer rains. Winter northeastern storms bring to the Cape snow and freezing rain. Over the course of a year the Cape normally experiences between forty and forty-four inches of precipitation, most of which runs through the highly porous sandy soil to an underground aquifer.

PLANTS ARRIVE, MARSHES FORM

As the ice retreated and the waters rose, the land warmed. Moss spread across the newly formed sandy land and seeds brought by wind and birds took root.

The original tundra plants, which came in the wake of the glacier's retreat, were followed by northern migrating species as the weather warmed two thousand years ago. Weather slowly cooled after that, but some southern species, such as white cedar, holly, inkberry, tupelo, broom, crowberry, golden aster, pinweed, jointweed, and meadow beauty, survived. These plants were able to keep their foothold on the Cape because of the warming effect of the ocean. Likewise, the summer cooling effect allowed such northern species as bearberry, heather, golden heather, leatherleaf, bunchberry, checkerberry, and water lobelia to maintain a presence on the Cape.

As plants spread across the ground, small animals migrated up the long peninsula. Where the water met the land, aquatic life flourished, most prolifically in tidal marshes. As the tides washed into shallow areas, they carried more force than when they receded. The greater force of the incoming tide brought in more sand and silt than the outflowing tide took out, which created marshes that rise with the rising sea levels. Cordgrass (*Spartina alterniflora*), tolerant of salt water, grew where the daily tides washed over the land. Salt meadow grass (*Spartina patens*), with a lower tolerance of salt, grew where only the highest tides washed. Cordgrass was a major building block in the developing marsh. Cold weather killed the cordgrass stems; as they broke off and fell to the marsh floor, the stems created marsh wracks, which washed about the water's edge spreading nutrients and organic matter. Eelgrass (*Zosterta marina*) in the lagoons, in the shallow-water estuaries, and in the bays supported a wide variety of marine life.

Marsh grasses provided the food and habitat for the insects that so bothered the guests of Lorenzo Dow Baker, but those insects, and the thousands of others that thrived in the edge area between land and sea, provided food for birds and aquatic life. Mice and voles, feeding upon grass stems, were themselves the food of foxes, weasels, otters, muskrats, and raccoons. The marshes not only supported a diverse community of fish that spent their lives in the marsh, but also acted as nurseries for larger fish who then ventured out to the wider sea. The marshlands also provided food and habitat for a plentitude of fowl.

TREES

As the Cape warmed, its land became green with grasses, bushes, and trees. Sandy, nutrient-poor soil proved suitable to the white pine whose forests spread out across the Cape between seven and ten thousand years ago. Some five to seven thousand years ago, as the needle litter grew on the forest floor, oak and hemlock began to crowd into the white pine forests. Hickory came to the Cape a couple of millennia ago. Within the last thousand years, chestnut, beech, ash, sour gum, and tupelo arrived. In wetter regions maple and yellow birch thrived, and in the low bogs, white cedar joined the peninsula's forest cover.

Over several thousand years, the cycles of life and death created a thick mass of organic ground cover. On the higher ground under the coniferous forest, where fewer burrowing animals lived, this plant litter created a thin layer of drained loam that lay on top of the sandy podzolic soil typical to coniferous forests. In the soil under the deciduous forests, which were more common inland along the valleys of the Outer Cape and on the north side of the ridge running down the Cape from Buzzards Bay to Orleans, burrowing animals and soil dwellers mixed the plant litter of the forest floor together with underlying layers of soil. The deciduous forest burrowers churned up the dark organic matter under the deciduous forests with the lighter, sandier soil below, creating a deep mix of loamy soil. Across the south side of the central Cape ridge was coniferous forest, much like that of the exposed areas of the Outer Cape, with a thin layer of loam resting on light and sandy soil. This light sandy soil held less water, and nutrients were easily leached out, but it did support thick coniferous forests that continued to supply organic material to the forest floor. In sheltered deciduous forest areas the organic loam, or humus, reached down to depths of two or three feet while in higher, more exposed coniferous forest areas the soil could be as shallow as a few inches. On the average, six to eight inches of topsoil covered the Cape. As in much of New England, because of the lack of a limestone base and the predominance of acid-producing deposits of oak leaf and pine needles, the Cape's soils are acidic. Shallow kettle holes filled in with organic matter and became lowland meadows while, in some deeper kettle holes that were also poorly drained, acidic peat bogs developed.

The yearly rains and snows of the Cape kept its water table high and prevented the invasion of ocean salt water. They also provided continuous flows of fresh water into the marshes, estuaries, and tidal inlets. In tidal marshes the mingling of fresh water and salt water allowed for the rich growth of grasses and brought nutrients into the food chain that supported shellfish, finfish, and shorebirds.

NOMADIC PEOPLES ARRIVE

Thus, it was a relatively young pine-covered Cape dotted with hemlock, oak, and hickory that greeted the first humans. Archeological evidence indicates that people first wandered on to the Cape during the Late Archaic period, three to eight thousand years ago, when the Cape was taking on its modern form. These were nomadic people looking for game and wild plants. These earliest explorers of the Cape found its shores teeming with life, particularly shellfish and finfish. These first inhabitants dug clams and captured fish and crabs. They trapped shore and marsh fowl and mammals and gathered wild berries. As the flora and fauna diversified across the peninsula, the Cape's human residents added larger animals to their hunting.

The sites that these first peoples left behind, scattered about the Cape, indicate transient use. Evidence of more human activity on the Cape comes from a time between the Early Woodland period, two to three thousand years ago, and the Middle Woodland period, one to two thousand years ago. These peoples constructed hearths for cooking and fashioned sharp projectiles. By the end of the Middle Woodland period, new technologies and tools and perhaps new people came to the Cape. These peoples used a greater array of projectiles and fashioned ceramic containers from shells, sand, and clay. Some of these they decorated with design.

Horticultural activity, notably the growing of corn, came to the Cape during the Middle Woodland period, but it did not come suddenly or uniformly. Some archeological sites, such as the Carns site at Nauset, give little evidence of either production or consumption of cultivated plants. Other sites, even in the Nauset area, indicate corn production. Native peoples of the Cape of some fifteen hundred years ago practiced a variety of land use strategies. Some, such as those who used the Carns site, primarily hunted and gathered moving from camp to camp. Others such as those in the Mashpee and Barnstable area cleared fields, planted and harvested crops, and gathered shellfish in abundant quantities. Those practices, rather than the mobile hunting and gathering of the people using the Carns site, were increasingly dominant, especially by the middle of the Late Woodland period.[3] Although the Cape shared a richness of fauna and flora with much of the rest of New England, the richness of its marshes and tidal flats allowed its peoples to settle down and develop the more permanent settlements that Europeans saw when they explored the Cape's coast four hundred years ago.[4]

A thousand years ago the peoples of the Cape girdled and burned trees in order to open fields. In mounds that they built with buried horseshoe crabs and fish, they planted maize, beans, and squash. The fresh fields allowed native peoples to spread across the land of the Cape. Woods and brush were burned, not only for planting but also to clear openings in the forests for travel and to create habitats for animals. Forest burning also facilitated hunting.[5]

By the start of the Late Woodland period, roughly two thousand years ago, the Cape's rich ecosystem allowed for both seasonal and large settled communities.[6] The mild weather along with the rich shellfish beds and the abundance of game animals and flora allowed the native populations of the Cape more permanent settlements than their mainland cousins. They no longer decamped each spring and fall. This encouraged the Cape residents to build large and substantial shelters. Even the winter storms aided the Cape's native communities. Strong winds and heavy surf stirred up shellfish, which washed along the shore for easy gathering. The huge mounds of shells and other waste found today in middens scattered about the Cape are relics of those settled communities. In order to clear underbrush, to limit more damaging fires, to decrease the number of pests (especially mosquitoes and black flies), and to open clearings for more diverse animal and plant life especially blueberries and huckleberries), the native population regularly set fire to the forests. Burning off the understory opened areas for hunting with bows and arrows. These periodic fires favored the larger slow-growing, thick-barked trees such as pitch pine (*Pinus rigida*), which dominated coastal forests from Cape Cod to Virginia. Along the coast the periodic burning combined with the strong onshore winds encouraged a mix of forests and open prairie grass. This provided the habitat for the eastern heath hen, an important food for those who lived by the sea.

Many of the hundreds of kettle ponds had outlets in the form of streams and creeks that flowed to Cape Cod Bay and Nantucket Sound. Each spring, as the opponents to Lorenzo Dow Baker knew and appreciated, hundreds of thousands of alewives and shad (both are members of the herring family), pushed up through those streams and creeks into the ponds to spawn.[7] The annual fish runs were vital to the survival of the Cape's indigenous peoples, providing sustenance in spring when other food supplies were short. Native communities caught these anadromous fish (i.e., those that migrate upstream from the sea to spawn) in nets or trapped them in weirs placed across the streams.[8]

But spring spawning runs were not the only source of herring for the Cape peoples. During the summer large, thick schools of menhaden entered the

Cape waters, as Henry David Thoreau noted in 1849: "The sea was spotted [with schools of] them far and wide."[9] Weirs were placed along the shores where these fish could be gathered and brought home for food and fertilizer for the mounds of planted maize, beans, and squash. And in the salt marshes of the Cape, gatherers collected grasses from which they wove mats and baskets and with which they covered their homes.[10]

Along with the fish and animals they caught and trapped, and the maize, beans, and squash they cultivated, Native communities also included in their diet the nutrient-rich plants that grew on the water's edge. They gathered beach peas and added them to their porridge of maize, beans, and squash. Sea rocket, scurvy grass, sea spinach, sea blite, seabeach orach, woody glasswort, and beach plums added vitamins, especially vitamin C, to their diet. On higher ground blueberries, blackberries, and huckleberries flourished. The availability of these edible plants near the water provided balance to the diet of the native population even in the late autumn. But it was the abundant shellfish of the Cape's tidal flats that enabled the Cape peoples to establish settled communities. Oysters were so prolific that huge reefs of them filled the Cape's estuaries. Simple tools and quick hands gathered hundreds of clams, mussels, and lobsters. The massive shell middens that are evident even today indicate the importance of shellfish for the diets of Cape Codders before the arrival of whites.[11] The wealth of the Cape's waters across the seasons, the long growing season, and the abundance of nutrient-rich wild plants at the water's edge combined to provide stable sources of food without the necessity of breaking up camp and moving on to other environments.

2
FIRE, FISHING, AND FARMING OF NATIVE PEOPLES

In 1605, the French explorer Samuel de Champlain dropped anchor in Nauset Harbor, where the Cape bends abruptly northward. He found large communities of several hundred Native Americans living in round "cabins," each "covered with heavy thatch made of reeds."[1] Cultivated fields of beans, maize, and squash surrounded the homes. Returning to the Cape in 1606, Champlain sailed into what is now known as Stage Harbor in Chatham, where he saw between five and six hundred natives living comfortably by the shore. Champlain described those homes as opening out onto large gardens of maize and squash. Around the gardens tall palisades protected the growing crops. Although Champlain noted that the Native Americans carried bows, arrows, and clubs, "they were not so much great hunters as good fishermen and tillers of the land."[2] That same year, in what would become Wellfleet, he visited the settled village of Punonakanit, where he noted that the villagers harvested shellfish in the winter, when they also hunted, and raised maize and squash the other months.[3] These villagers were members of what would be known as the Wampanoag Indians.[4]

NATIVE AMERICAN ECONOMY

Both the Cape's soil and its climate—persistent mild rains, sun, and a long growing season—encouraged horticulture. The Cape's sandy soil could easily be worked with wooden and shell digging tools. By using horseshoe crabs and fish to fertilize their cultivated mounds, these Cape residents were able to produce enough food to sustain large stable communities.[5] Cape Native Americans wove watertight grass baskets from marsh grasses, filled these baskets with the harvested crops, and buried the baskets several feet under the sand. Such stored provisions provided food (and planting seeds) over the late winter and early spring, thereby supplementing the clams and oysters harvested from the coast.[6] The rich deposits of clay left behind by the glaciers, and exposed in seams along the face of coastal dunes by wave erosion, provided material to make ceramic pots to store their harvest and protect it from moisture and pests.

Although Champlain found large settlements with hundreds of residents and mounds of waste shells, he did not see a wasted land. We do not know how many Cape communities may have failed to develop a sustainable relationship with the land over the several thousand years humans interacted with the Cape environment. But the Native Americans, who built dirt mounds to plant their maize, beans, and squash and harvested shellfish and finfish from the sea, appear to have managed their natural world. They opened clearings in the forests for cultivated fields and villages and built palisades around their cultivated fields to protect them from raccoons, bears, and deer. They burned brush to open up clearings in the forest and allow grasses and new fruit-bearing bushes to grow. This created edge environments that encouraged heath hens, pigeons, deer, and rabbits and facilitated their capture. They constructed weirs across streams and along the shore. Still, aside from their forest burnings, they scarcely reorganized the Cape's ecosystem. Altogether the Native Americans actively cultivated between 1 and 2 percent of Cape land. Approximately another 2 percent lay in abandoned cornfields slowly returning to forest. More than half the Cape was forest that was subjected to periodic burning by Native Americans. One percent of the Cape was forested in white cedar. The rest was either dense unburned forest or swamp and marsh. Many of the trees—oak, pine, and beech, even those subjected to burning—were massive with diameters of fifty inches or more.[7]

Trails were established linking the various coastal settlements.[8] One trail ran from Eastham to Yarmouth linking the Nausets with the Mattakeese. Another linked the Mattakeese of Yarmouth to the Cummaquids in Hyannis. The communities in Yarmouth Port and Hyannis were joined by a trail to the Manomets at Sandwich. These trails were so well worn that white settlers incorporated them into their initial roadways that today carry automobile traffic across macadamized highways.[9]

Henry David Thoreau reminded us when he visited the Cape two and a half centuries after Champlain that "at every step we make an impression on the Cape, though we are not aware of it."[10] Yet the "step," or as we would call it today, the "footprint," left by the Native Americans on the Cape, was a small one with a soft sole.[11]

PLAGUES

Champlain was not the first European to touch the Atlantic shore.[12] Bartholomew Gosnold explored the Cape in 1602, and others had come even earlier. Explorers sailing into what they thought were seas unknown to Europeans in 1583 discovered thirty-six English, French, Spanish, and Portuguese vessels fishing off the coast of Newfoundland. Fishermen also worked the banks off the coast of Nova Scotia. The fishermen found refuge from Atlantic storms, resupplies of fresh water, and a place to dry fish on the shore, while traders came looking for furs; they all observed and interacted with those for whom Cape Cod was home. These visitors took away water, food, furs, and stories, but they left behind death.

While the girl in the maize field was chasing away crows, a foe far more threatening was moving across Cape Cod. Between the time when Champlain sailed into Nauset Harbor and the Pilgrims had their "first encounter" with the natives of Cape Cod in 1620, disease—most probably smallpox or possibly a particularly viral typhus, trichinosis, or even bubonic plague—swept through the native tribes.[13] For some nine thousand years, due to the rising seas of glacial melting, the peoples of the Americas had been cut off from the rest of Eurasia. They had no domesticated animals, so they did not live in close enough proximity with animals to allow diseases to move between animals and people. Europeans, with their wide trade contacts, had centuries upon centuries to develop a living relationship with a varied disease pool. The peoples of the Americas, on the other hand, were relatively disease-free. But disease-free was not the same thing as disease-protected. Even the childhood diseases of Europeans proved lethal. Death rates, in proportions far greater than the great plagues of fourteenth-century Europe, swept through Native American communities. Whole villages were destroyed leaving survivors too weak to hunt or plant.[14]

When the *Mayflower* sought refuge on the bay side of Cape Cod, the expedition party that included Myles Standish and William Bradford did not find prosperous native villages. When they ventured ashore in present day Truro, although they found stashes of food (which they proceeded to steal, while claiming that they intended to pay for it), they found few people. In Eastham while ashore looking for water, the Pilgrim party encountered a group of Native Americans. These native peoples were less friendly than the ones who greeted Champlain. Perhaps the natives who confronted the Pilgrims with arrows were angry because the Pilgrims had pilfered their food. Perhaps they remembered the earlier visit of the Englishman Thomas Hunt, who had responded to the friendship offered him by capturing and taking off natives to sell as slaves. Or perhaps they were simply terrified to suddenly realize that contact with Europeans meant death.

And death continued to follow the Europeans. Waves of smallpox returned again and again even after the Pilgrims established themselves at Plymouth. These epidemics took a toll on the European settlers, but they had a far greater impact on the native peoples. And disease was not the only destructive force visited upon the original inhabitants of New England by the Europeans. The superior weapons and the disruptive force of European capitalist markets also played havoc on the communities and traditions of the native peoples. The Dutch, the French, the Pilgrims, and later the Puritans all pushed European goods, pots, pans, knives, blankets, and coats in exchange for furs—above all, beaver—and land.[15] Although these European goods were commonplace in Europe, on this side of the Atlantic they were held in high esteem. Hunting for furs to exchange for European goods soon replaced hunting for subsistence, dramatically increasing the pressure on local fauna. Failing to realize the full extent of the European sense of exclusivity of property, and having a healthy respect for the lethal power of European weapons (and white peoples' proclivity for using them against native peoples), local sachems found themselves selling off so much land that their people's way of life was jeopardized.

After a century of European contact the Cape survivors lost the independence they once enjoyed. When Champlain first visited, over two thousand native peoples lived off the land and sea of Cape Cod. By the middle of the eighteenth century fewer than five hundred remained. The huge Monomoyick village Champlain saw in 1606 was reduced to five or six families a century later. Most of those native peoples who survived on the Cape lived hand to mouth, selling baskets, clams, or kindling to local white families. The only native community of any size was Mashpee, which survived through white guardianship.[16]

II
THE ERA OF LOCAL RESOURCE PRODUCTION AND EXTRACTION
Settlement to the Start of the Twentieth Century

3
ON THE WAY TO AN AMPHIBIOUS SOCIETY

When the *Mayflower* sailed around the northern tip of Cape Cod and dropped anchor in its well-protected hook, the hardy zealots who had left England after their sojourn in the Netherlands were pleased their God had finally delivered them from the wilds of the North Atlantic. It had been a long time since they had last seen mayflowers or greenery of any kind, but on this November day they looked out upon a sylvan land. For a people who had lived for several years in Leiden with few trees, the forests just inland from the coast of Cape Cod, even in winter, looked thick and lush. Pilgrim diaries noted that the Cape was wooded in oak, pine, sassafras, juniper, birch, holly, ash, and walnut in forests "for the most part open and without underwood."[1]

Although they were impressed with the woods of Cape Cod, the Pilgrims were concerned about fresh water supplies and the quality of soil they found on its tip. After a stay of five weeks on that wild but protected tip of Cape Cod, during which they drew up the famous Mayflower Compact, they decided to push on to Plymouth Harbor.

Arriving in Plymouth in late December of 1620, the small party began building shelters. The first, a twenty-foot-square common house, was finished in early January. Seven more houses went up that first year. In the meantime, most settlers spent the winter on the ship. It was a hard winter. Half of the original 106 who set sail from Plymouth, England, failed to survive until summer. By spring, the survivors were ready to plant. In April of 1621, Squanto, a Native American conversant in English (he had twice been kidnapped and had spent several years in England), greeted them. Surprisingly friendly considering his original treatment by Europeans, he taught the settlers how to grow maize, beans, and squash in this new land. Although Squanto's help enabled them to begin growing food, it was not the intention of these settlers to live by the customs of the Native Americans. They were Europeans, and they hoped to convert this Indian-modified wilderness into a European agricultural community.[2] To accomplish this, they had brought seeds, chickens, goats, and pigs from home. In 1623, William Bradford ended the practice of collective fields and began the process of allocating lots (for planting, pasture, and meadow) to individual families. Each new boatload of Pilgrims arriving from England brought more seeds and more animals. By 1636, cows, then more horses and cattle, arrived, and with this livestock the thick meadow grasses took on importance.

Although the Pilgrims' plans involved transforming their new home into something that resembled their old home but with much more godliness and greater dispersal of land ownership, the physical world they hoped to transform was different from the one they left behind. Winters were significantly colder and longer than in England. Pastures turned green far later in the spring and died back much earlier in the fall. On the coast, winters were not only colder but stormier as well. Northeastern storms pounded the Pilgrim settlement throughout the winter while the summer occasionally brought epic storms that left near-total destruction in their wake. The young community was only fifteen years old when a ferocious hurricane swept through blowing down whole forests.[3] Although the colonists put the downed trees to use by cutting them up for cordwood or lumber, the storm impressed upon the settlers the power of nature.

In New England the growing season for grains was short, and the soil was sandier and held fewer nutrients than the rich soils of England. Still, although the sandy soil was not as fertile as the clayey loam of England, it was easy to work, and it drained well. With plenty of rain (42–44 inches a year) and the addition of some form of fertilizer, it made excellent planting ground for maize, beans, and squash, as well as for the English crops of barley and rye.

Native Americans dealt with the tendency of nutrients to leach out of sandy soil by deeply burying fish or horseshoe crabs in their planting mounds. The English settlers preferred shallow planting across a broad, level field. Dead fish or crabs scattered in such fields drew animal scavengers that would dig up the fields and destroy the plantings. The Pilgrims, instead, used manure from their domesticated animals: horses, cows, pigs, sheep, and chickens. Manure could fertilize the sandy fields, and it held little attraction for scavenging animals. But livestock also had to be fed.

In England farm animals could be pastured for most of the year, but in New World Plymouth livestock had to be brought in out of the winter weather and fed while their pastures were bare.[4] The Pilgrims soon learned that wild salt marsh hay, which grew in abundance along the high marsh, made for excellent winter fodder. While their animals were overwintering in protected shelters and eating marsh grass, manure accumulated. Come spring, the Pilgrims spread the manure across their grain fields before planting.

By the time of the great hurricane of 1635, the settlement in Plymouth had grown large enough that it was pushing against its territorial limits.[5] To the north, Plymouth settlers filled in the land of what became Duxbury and Marshfield. Further north, the new colony of the Massachusetts Bay Company established itself on the better land around the deep harbor of Boston. Looking south to Cape Cod, the Pilgrims saw a land of woods, good soil, and plenty of marsh grass, an important consideration given the rapidly growing number of cattle. The Cape's great meadows, especially those in what became the town of Barnstable, were already known to these early farmers. Plymouth residents had been sailing down the coast to the Cape to cut and haul back marsh hay for animal feed before they decided to set up settlements on the Cape. Thomas Prence, a future governor of the colony, noted in 1644 that the soil of the Cape was the "richest soyle for ye most part blakish and deep mould."[6] Another Pilgrim visitor to the Cape claimed it had "excellent black earth."[7] This rich soil existed under the canopy of a thick deciduous forest that spread out to the north from the ridgeline that ran down the center of the Cape as well as along the inland/western side of what became Eastham. Coniferous forest covered much of the rest of the Cape inward from the shore.[8]

CAPE SETTLEMENT ALONG SHAWME POND

In 1637, a group of settlers from Saugus in conflict with the authorities of the Massachusetts Bay Colony asked the Plymouth government for a grant of land south of Plymouth. It lay just over the narrow land that connected the

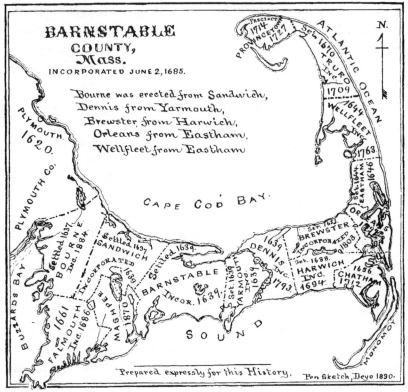

Map of Cape Towns and Years of Incorporation, from Simeon L. Deyo, History of Barnstable County, Massachusetts *(1890).*

Cape to the mainland, in what is today Sandwich and Bourne. Given the grant, a group of ten families began clearing forest for the new settlement in 1638 along Shawme Pond. The settlers brought their livestock with them. The soil was rich and loamy, and Duck Creek and its marshes were only a short distance away. Within four years, fifty other families joined the initial settlement. Sandwich appealed particularly to these settlers because of their need for pasture land and hay for overwintering their stock.⁹ They let their pigs loose to run about the woods. Despite some dangers from wolves and other predators, the free-running pigs thrived. On the other hand, pigs could root up and destroy planted fields and gardens. As Richard Bourne discovered in 1638, his first year on the Cape, the Plymouth court required all pigs to have rings in their snouts to discourage rooting. Failure to ring one's pigs brought a fine and a trip to Plymouth court. Bourne was not alone. Out of the small number of families settling in Sandwich within a year of settlement ten people were fined for failure to ring their pigs.[10]

The community kept a closer watch over its horses and cattle than over its pigs. Pigs could fend for themselves for most of the year, but horses and cattle needed to be fed over five months a year. For this the settlers needed marsh hay, and the marshes were not so abundant that the community could let anyone cut at will. During the first ten years of settlement even freeholders were required to gain permission from the court to "cut hay at the ponds beyond Sandwich plains."[11] Marsh hay provided fodder for animals over winter, but cutting hay for year-round livestock feed would have been far too labor-intensive. Once spring arrived, farmers turned their livestock out to pasture—often dune grasses and drier meadows. As in the case of marsh hay, the town found it needed to manage its pastureland.

By 1654 Sandwich had grown to over threescore families, almost all of them with cattle they needed to pasture. Since the first years of settlement, farmers had put their cattle on the commons of "Town Neck," a large land spit east of the town harbor between the bay and Old Harbor Creek.[12] Grasses grew "luxuriant" on this sandy peninsula where winds kept down competing trees, and the cattle grazing on the spit were more easily protected from predators. But overgrazing began to concern town leaders. In 1652, the town restricted grazing on the Town Neck from May 1 to October 4, and stipulated, "No cattle except calves shall be put in without the consent of the town."[13]

Cattle were not the only animals grazing upon common land. Horses came with the original families, and, within a few years, sheep too had grown plentiful. Sheep provided wool, an important commodity both for trade and for spinning and weaving into cloth for blankets and clothing. Sheep did well on the Cape. The winters were milder than on the mainland and the wild grasses and brambles of the meadows provided plenty of food. But unlike the pigs, sheep could not take care of themselves. Predators, especially wolves, migrating in from the thick woods of the south and west, exacted a heavy toll on the flocks. To reduce the killing of their sheep, the town offered a bounty of between five shillings and one pound, depending on their size, for the scalps of wolves. In a community with little liquid money, the bounty for wolf scalps was high. One could afford to build a home with a bounty from a couple of large wolves.

Managing the town's grasses, controlling its pigs, and killing off wolves were not the only problems confronting these early proprietors. Although many of the original families had come from Saugus, the town of Sandwich was part of Plymouth, which lay twelve miles to the north on an old Native American path through heavy wood and loose sandy soil. In 1652 Plymouth charged the town leaders to lay out a road linking Sandwich to Plymouth. But completing a passable road proved difficult. It was far easier to sail between the two settlements. For much of the next two centuries that is what the residents did.

Fall and winter storms made both overland and sailing trips arduous; yet those were the seasons when the farmers of Sandwich had harvested grain to grind, and the mills of Plymouth lay in the distance. In 1652 Sandwich began negotiating with Thomas Dexter to build a mill convenient to town farmers. These initial negotiations failed to produce a mill. A new committee of town leaders paid John Ellis, William Swift, William Allen, and James Skeff twenty pounds to build a mill, which was completed in 1655. They engaged Mathew Allen to "grind grain and have a toll for his pains."[14]

Allen's mill and the later ones built by Dexter and the town itself answered the problem of getting grain milled without having to travel up to Plymouth, but the dams built to create the waterpower necessary to turn a heavy millstone inhibited the spawning path for herring and other anadromous fish. And it soon became apparent to the settlers how important these fish were.

BARNSTABLE AND THE GREAT SALT MARSH

Two years after the Sandwich settlers began building homes on the northwestern edge of Cape Cod, a small group from the northern section of the Plymouth Colony established a settlement roughly ten miles east of Sandwich at Barnstable, a land with an immense marsh and good loamy soil. Conveniently enough, Native Americans had earlier cleared the planting grounds, and they had not yet become overgrown with forest. The settlers favored Barnstable, not only because of its immense marsh and good soil, but also because its "forests were filled with wild game, the bay was filled with fish and at low tide . . . [and] clams and oysters and schools of alewives" were abundant.[15]

Within five years of the first settlement, the new town of Barnstable designated a local sachem as owner and persuaded him to sign over several thousand acres of land to the south of the settlement in exchange for four coats and three axes. The natives believed that such an exchange did not preclude them from fishing, hunting, or gathering wood on the land they had sold. At this point the settlers were not interested in these details, which unfortunately would present problems later. Getting access to the land to distribute to newcomers seemed the more important concern. A second and third purchase, from other sachems, cost the town two more coats and two brass kettles and gained the settlers a substantial tract of land stretching from Cape Cod Bay to Nantucket Sound.[16] These purchases gave the town more land than could be cultivated or fenced. Most of the land therefore was left open as "commons for the town's cattle."[17]

THE ROCKY SOIL OF PLYMOUTH

While the settlers were establishing their fields and husbanding their animals in Sandwich and Barnstable, the Pilgrims of Plymouth began to find their settlement not as much "God's country" as they had initially presumed. Plymouth's soil was hard to work. It shared with most of New England an acid soil and an abundance of rocks. The glacier that plowed across the Appalachian Mountain range on its way to Cape Cod picked up more than sand and gravel. It also took up stones and boulders, but it distributed them unevenly. Some it left behind on Cape Cod, as the Sandwich and Barnstable pioneers could attest, but a larger proportion of big ones settled inland from the Cape. Because it was land created mostly from the outwash end of the glaciers, the Cape was more heavily favored with sand and gravels, and less so with boulders. Inland, the glaciers had scraped off surface soil and left behind an abundance of large stones, rocks, and clay; "a scraggy scoria of a country," Melville's Ishmael called it.[18]

In the thick forest surrounding Plymouth Harbor, the Pilgrims found wood for homes, for fuel, and even for export back home to England. They did coax maize and beans as well as European rye and some wheat from their land, but they were discouraged by the small returns from their fields. Indeed, for several years the colonists sailed down to the territory of the Nausets at the eastern edge of the Cape to trade cloth and trinkets for the Native Americans' surplus maize.

In 1640, after twenty years, many in Plymouth became convinced that they could not build a capital of godly people on such barren soil. In 1643 a committee of seven, including Governor Bradford, traveled to Nauset to assess its feasibility as a new center for the church and government. Although the committee reported back that Nauset seemed too far from the rest of the settlement to function as the government and religious center, it decided to allow some Plymouth families to establish a new plantation there, and it granted the settlers land covering the whole area from what is today Wellfleet to Monomoy Point. The new settlers then purchased land rights from native sachems. They reserved to the indigenous people the rights to shellfish and a share in the blubber of beached whales, as well as reserves of land such as Indian Neck in Wellfleet.

By 1651, a new town with seven families of some fifty residents was incorporated and renamed Eastham. Each of the original settlers took up farms of about two hundred acres composed of marshland, woodland, rich and easily worked sandy plow land, and grass-covered dunes. The settlers let their cattle roam the grassy dunes over the spring and summer.[19] In the fall at the low ebb of the spring tide, when the water recedes from the marshes, these farmers harvested the salt hay (marsh grass), hauled it to their barns, and gathered their animals from the dunes to overwinter in sheltered pens.[20] In the early spring they mucked out their barns and pens and hauled the manure to their planting fields. These they plowed and seeded with wheat, maize, rye, and oats. Production from their fertilized fields proved sufficient to feed the families of Eastham and provided a significant surplus for export.

Instead of using indigenous planting practices, the white settlers scattered their seeds over freshly plowed fields, which put the farmers in constant conflict with birds. The settlers responded, as they did across New England, by attempting to eliminate the birds. Each housekeeper was required to kill twelve blackbirds and three crows a year; each single male resident had a quota of three crows and six blackbirds and was not allowed to marry without producing this quota of dead birds.[21] Eastham's productive fields soon demanded mills to grind grain. Two small watermills were built in 1683, with larger windmills soon to follow in the next century.[22]

EIGHTEENTH-CENTURY FARMING

By the end of the seventeenth century the Cape was well settled with white people. Villages and farms stretched from its western edge up to the sandy lands that reached out to the bay at its northern tip. The settlements were composed of several essential farm elements: woodlots, marshland, pastures, and planting ground. A typical Cape farmer worked thirty-one acres with the rest of his sixty-to-seventy-acre farm in marsh or woodlot. Farmers attempted to hold diverse types of land: sandy soil for plowing and planting, heavy lowland soil for orchards, upland soil for pasture and woodlots, and marsh or meadow land for hay. Forests were cleared for planting crops, and trees were cut and sawed to build the squat little homes that later took their name, "the Cape Codder," from their place of origin.[23] Initially, the Cape was heavily forested, and the cutting facilitated the opening up of farmland. Trees were cut and stumps pulled to make way for plowing. Settlers cut up some trees for cordwood and used other trees for construction, notching wood for house and barn beams, cutting it for floorboards and siding, and splitting it for roofing shingles and shakes. The demand for open land for plowed fields, as well as the need for wood for fuel and building material, began to eat away at the Cape's dense forests.

These homes could be built with local lumber (oak) for twenty-five dollars in the early eighteenth century. By contrast a two-story oak-framed house with a stone chimney would cost one hundred dollars to construct using hired labor to cut the lumber and purchased materials not available on the land.[24] Timothy Dwight, a well-known Congregationalist theologian and president of Yale from 1795 to 1817, traveled across the Cape at the beginning of the nineteenth century and commented on the utility of Cape Codders' homes. They were, he noted, "usually one story with four rooms on the lower floor and covered on the sides and roof with pine or white cedar shakes. The chimney is in the middle."[25] These homes were low to the ground, with small windows and low ceilings. Their shingled sidings protected them against the harsh winter winds and summer storms. The central fireplaces, however, consumed a great deal of wood. A typical Cape Cod home burned between twenty and forty cords of wood a year. In the early eighteenth century the town of Chatham provided its minister with thirty-six cords, sixteen of oak and twenty of pine.[26] The storage space for the wood delivered to the Chatham minister would have filled up a four-foot-high space, sixty by forty feet. The annual consumption of twenty to forty cords of wood per family took a heavy toll on the Cape's forests. With each acre of forest producing thirty cords of wood, an ordinary farm needed some thirty to forty acres of woodlots to sustain a yearly harvest of wood. Increasingly Cape settlers became concerned about the availability of wood. By the second half of the eighteenth century, allotments of wood to local ministers declined as cordwood became scarcer.[27]

By 1712 Truro elders declared "that in consequence of great waste being made of wood in burning lime to be sent out of town which may cause a scarcity of firewood, no persons must cut on the commons for that purpose." The town also voted against wasteful "cutting of cordwood and timber."[28]

MIXED FARMING

After clearing woods, farmers plowed fields and sowed wheat, rye, and maize.[29] Close to their homes, which they located in valleys to find shelter from the punishing winds that swept across the Cape in the winter months, they put in garden plots of cabbages, turnips, carrots, beans, and peas. In low areas at the edge of wetlands, where the soil was thick and clay-laden, they planted orchards.[30] They preferred apples, but by the early eighteenth century they were planting pears, which were more tolerant of the wet soil. In addition to their pigs, cows, and horses they bred sheep for wool. Because at the end of the seventeenth century the population of Massachusetts Bay Colony to the north around Boston grew faster than its ability to feed itself, cattle raising became profitable on the Cape. All of the Plymouth Colony, and especially the western Cape towns, shipped and drove local livestock to Boston markets. Eastern Cape Codders grazed some ten thousand sheep in 1660. By the eighteenth century the Cape was a major producer of wool and mutton.[31] Most of these sheep, cattle, and horses grazed on beach grasses during the summer.[32] Although this practice reduced the need of individual farmers to maintain their pastures, it took its toll. In 1730 Truro appointed a committee to "prevent cattle and horses going upon the meadows and beaches adjoining" in order to prevent the destruction of the meadows by blowing sand.[33]

As these early settlers raised more grains and sheared more sheep, they required more mills. In response, towns granted land at the edge of streams for the construction of gristmills and sawmills as well as the fulling mills where wool could be pounded and washed. Town grants required their milldams to contain facing boards that could be raised in the spring to allow fish to freely migrate to and from spawning grounds.[34] Where streams were meager but wind plentiful, towns appropriated funds for the construction of windmills. The farmers paid the miller a portion of the grain for grinding.[35]

AN AMPHIBIOUS CULTURE AND ITS DEMANDS

The initial settlers on Cape Cod were farmers, and they had come to the Cape to take advantage of its rich loam and lush meadow grasses. Water, however, surrounded the clusters of settlements that spread across the Cape's sixty-mile length. The need to move goods to mainland markets pushed Cape Codders to the sea. After three generations of settlement, Cape Cod emerged into an amphibious culture. Boats moved people and goods between towns as well as on and off the Cape. The sea not only provided an avenue to move goods, but, more significantly, it became the source of goods. Since the earliest settlements, the farmers of the Cape had used the fruits of the sea to supplement their agriculture. Clams, lobsters, and oysters quickly entered their diet, and spring herring runs added protein to the settlers' meals, as they had for the Native Americans before them.

Oysters could be exported. In the eighteenth century Wellfleet became famous for its oysters, which it shipped in barrels to markets in Boston and New York. Unfortunately Wellfleet oystermen overharvested the beds, and local farmers smashed up the shells to spread on their fields. No one considered that oyster shells needed to be returned to the bay because it was upon those shells that seed oysters took root and grew. Thus Wellfleet's oyster colony fell prey to disease, which all but wiped it out.

John Smith, who explored these waters in 1614, prophesied that produce of the sea, especially cod, would be the wealth of the region. The early Pilgrims were farmers, not sailors or fishermen, unable to act on John Smith's foresight. Yet the sea, for these early settlers, had a way of dumping riches on their shores. In the late fall, winter, and early spring, small whales (*Cetacea*), usually pilot whales (which the settlers called "grampus"), cruised the bayside shores. The shallow tidal flats that lined the bay from Provincetown to Brewster proved particularly dangerous for these large marine mammals.[36] Almost yearly, schools of whales stranded in shallow water by the outgoing tides served up a bounty unlike any harvested from the land. Native Americans prized the whale fat for cooking and eating, whereas colonists boiled the fat in large vats to make oil for lighting. As early as 1652, the towns appointed committees to adjudicate the disposal of stranded whales, dividing up the proceeds between the towns and any natives who may have first come upon the whales.[37] This profitable activity soon drew the attention of Plymouth, and the General Court (the governing body of the colony) ordered that one barrel of oil per whale be allocated to the colony. The oil commanded a significant price, which in turn the colonial communities could allocate to pay ministers and schoolteachers.[38]

Success at "trying" (boiling down) the blubber of stranded whales encouraged Cape Codders to become more proactive in bringing whales to the shore. By the end of the seventeenth century, lookout stations and watches were set up on high points overlooking the bay. When whales were spotted, boats put

Large marine mammals, such as these pilot whales, regularly get stranded on Cape beaches, sometimes in great numbers. The blubber of these stranded mammals was boiled down into oil, which provided an important source of income to the Cape. Photo courtesy of the Library of Congress.

out from shore and drove the whales onto the beaches where they were slaughtered. Larger whales were harpooned at sea and towed back to the shore for trying. In 1755 a Truro town meeting to discuss the vital issue of hiring a new minister was adjourned when news spread that a whale had been spotted in the bay and "so many of the inhabitants [were] called away from the meeting by the news of the whale."[39] By the 1770s over one thousand whaling ships sailed out of Cape harbors looking for whales in the North Atlantic. In 1771 Wellfleet had thirty whaling ships with total tonnage of 2,600 and employing 420 men. These whalers brought home more than 4,200 barrels of whale oil a year.[40]

Boatbuilding

The surpluses that accumulated at millers' storehouses and in farmers' barns represented the growing wealth of these new communities. But to realize their potential value, these surpluses needed to reach larger markets where they could be exchanged for other commodities, such as tools necessary to make the land productive for settlers accustomed to established ways. Those markets were initially Plymouth, and then Boston, and ultimately the larger Atlantic world. Sailing, not traversing difficult roads, proved the simplest and quickest way to move bulk items from the Cape to the markets to the north. The demand for seagoing vessels for trade and whaling moved Cape farmers to cut up, dig out, and shape wood from their forests into boats. Local oak, pine, and ash provided planking, framing, booms, and masts for vessels capable of negotiating the often-rough waters of Cape Cod Bay. Unlike houses and barns, which require straight, right-angled pieces of wood, boat builders needed curved wood. To keep the strength over a long curve, wood used for bow stems, knees, and ribs had to have the grain of the wood flow through the bend. These pieces were found where the roots or branches of trees curved off from the trunk. Boat planks strong enough to withstand the pounding of waves had to be flexible enough to bend across the broad beam from the stern to the bow. The Cape's rich seventeenth-century forests provided the needed wood for these vessels.[41] Along the water's edge, from Buzzards Bay to Cape Cod Harbor at the Cape's northern tip, woodworkers began specializing in building coasting boats that could carry Cape Codders and their produce off Cape and bring back what the Cape could not produce.

Cape shipwrights began by building dories and small coasting vessels and repairing larger boats. By the eighteenth century, beginning with the establishment of Thomas Agrey's shipbuilding works, they built schooners with the holding capacity of thirty to one hundred tons (in volume), which were capable of navigating the open Atlantic.[42] Shipyards sprang up along the coast from Sandwich to Provincetown for the production of small dories as well as the construction of schooners destined for distant fishing banks. The typical eighteenth-century fishing vessel was a round-bottom, two-masted schooner of between forty and (by the early nineteenth century), seventy tons.

Ship builders preferred white pine for planking, masts, and booms, and oak, ash, maple, and (by the early nineteenth century), locust for ribbing, keels, and stems. Shipwrights sought pine and white oak from the Cape's thick forests. These they supplemented with lumber imported from the coast of northern New England.[43] Although fire-resistant pitch pine covered large huge swaths of Cape land, its wood, strong and easy to work, did not hold fasteners as well as either the white pine or the hardwoods, a concern for ships in the rough Atlantic. Because iron rusted out when exposed to seawater, planks were attached to ribbing with wooden pegs that expanded when wet. These pegs held pitch pine planking, which reduced, but did not eliminate, the need of builders to hunt for wood supplies off-Cape.

The boats built and repaired in Cape boat- and ship-works needed tackle and fasteners made from wood, which also came from the surrounding forests. The many wharves that dominated the Cape shoreline by the end of the eighteenth century also used local timber in their construction. Next to them stood warehouses, sail-works, ropewalks, stores, and blacksmith shops, which utilized local charcoal to fire their forges.[44] By the middle of the eighteenth century the region's forests of white pine, white oak, ash, and chestnut had been depleted. The regrowth consisted primarily of pitch pine plus black and bur oak.

A New Fishing Economy and Its Needs

In this new century fishing grew to dominate the Cape's economy. By the 1730s, Cape fishermen were sailing beyond the local fishing grounds to the Grand Banks to compete for cod with fishermen from Nova Scotia, Newfoundland, and northern New England. Some fishermen dried their catch on local beaches before sailing down to the West Indies with their fish. They returned with sugar, rum, and spices. Others salted down their fish at sea and brought it home to the Cape for drying and further salting.[45]

By the last decades of the eighteenth century more than 665 fishing boats left Cape harbors for the fishing grounds. Chatham had more than thirty ships of thirty tons each, employing a total of 240 fishermen catching 12,000 quintals of fish.[46] Provincetown had four vessels (each of forty tons), with a total of thirty-two fishermen bringing home 1,600 quintals of cod. Neighboring Truro sent out eighty men in ten boats for 4,000 quintals of fish, whereas Wellfleet sent out three vessels with twenty-one onboard for a catch of 900 quintals. Along the south shore, Yarmouth had thirty vessels of thirty tons with 180 onboard and pulled in 9,000 quintals.[47] Getting their catch to market required that the fish be preserved. Fish had to be salted so it would keep until the ship reached its homeport. Once home, the drying process required more salt.

Salt was initially imported, then supplemented with locally produced sea salt. Salt makers created sea salt by boiling down salt water in large kettles. This consumed large amounts of wood. The production of a batch of ten bushels of salt in these large kettles consumed a cord of wood. In the late 1770s, when imported salt sources dried up due to conflict with England, the Sears family of Dennis experimented with solar evaporation for salt, which reduced but did not eliminate the need for wood. A typical fishing schooner required dozens of bushels of salt for each voyage. More wood still went into making the watertight containers for the salt and barrels for packing the fish.

By the mid-eighteenth century, the Cape was thoroughly an amphibious society. Boatbuilders needed wood not only for building their vessels but also for the resin they used to caulk them and to protect rope from seawater. Cape Codders harvested resin from local pine trees. Although it is possible to harvest resin from pine without destroying the trees, carelessness and haste often led to the killing of the trees in the process.

ADDITIONAL DEMANDS ON FORESTS: Causes and Effects

The use of wood as a fuel source (for heating homes and salt production) and as a building material (for houses, barns, fences, and sailing vessels) was not the only pressure on the Cape's forests. Barnstable's brick-works, which supplied the bricks upon which Cape Codders rested their homes and enclosed their chimneys, fired the bricks with local cordwood. The Cape ironworks utilized bog iron and smelted it with charcoal made from local wood.[48] These demands pressed in as well upon the region's forests.

Although extensive forests remained in the western Cape towns throughout the eighteenth and early nineteenth centuries, the eastern Cape faced serious shortages. Already by the end of the seventeenth century, Outer Cape communities were taking action to protect their wood supplies. In 1690 Eastham prohibited the shipping out of town of any wood cut from the town commons. Four years later, the town outlawed the exporting from the town of *any* wood, no matter where it was cut, and in 1695 the town prohibited the cutting of wood for any purpose from the town commons because "great damage hath been done."[49] By the eighteenth century, farm families in these towns turned to driftwood and peat to warm their home and cook their meals.

The cutting of forests from the lands of the Outer Cape had an impact that reached down almost to the shoreline. There, where strong winds blew off the water, short stunted trees and bushes grew near the beach along with beach and poverty grasses. These bushes and trees had little value as a building material, but they were used for kindling and for building "flakes," or fish drying racks. Along the beaches and behind fishermen's homes were acres and acres of such fish flakes. The removal of stunted trees and bushes, the cutting down of the forests just inland from the coast, and the practice of letting animals graze on beach grasses of the dunes (especially sheep, which not only ate the surface grass but nibbled down to the roots) loosened the already unstable sandy lands along the water's edge.

Erosion and blowing sands compromised harbors. Provincetown Harbor, called the Cape Cod Harbor in the eighteenth century, was the first to suffer significant sand erosion. In the early 1700s, sand began to fill in the harbor. In 1714 the Massachusetts General Court, the legislative body whose jurisdiction included the former Plymouth Colony as well as the Boston-centered Massachusetts Bay Colony, expressed concern about the Cape. Noting that the harbor was made "wholly unserviceable by destroying the trees . . . being of great service to keep the sand from being driven into the harbor by the wind," the General Court passed "An Act for Preserving the Harbor at Cape Cod." The act prohibited people from "bark[ing] or box[ing] any pine tree or trees."[50]

Because of the importance of the harbor at Provincetown and because of the increased sand deposits, the General Court returned more than once to the problem of sand erosion. In 1729 the court became concerned, not only about the cutting of trees and bushes, but also about the practice of grazing cattle on the dunes. "Whereas many persons . . . frequently drive down great numbers of neat cattle and horsekind to feed thereon, the beaches there are much broken and damnified which occasions the sands blowing into Cape Cod Harbor."[51] Provincetown, however, was not the Cape's only town struggling to protect itself from land abuse. In 1743 Harwich passed an order to protect its beach meadows and sedge grounds by outlawing "cattle, sheep, etc." on the grounds between Eastham and Mill Creek in Harwich.[52] Even Barnstable, which in the seventeenth century had seemed to have immense forests, asked the new state government in 1785 for legislation to prohibit the cutting of wood on Sandy Neck in order to protect the meadows from the inroads of drifting sands.[53]

Throughout the eighteenth century, town and colonial (and then state) governments passed a series of bills and laws restricting land use. Cutting of wood and boxing of trees were controlled, limited, or outlawed.[54] Governments passed legislation protecting dunes and limiting grazing. The town of Eastham, for example, passed ordinances limiting not only the cutting of trees but also the cutting of "thatch or sedge" before mid-August. In 1700 the town appointed Samuel Mayo and John Doane "to prosecute to effect in law against any person or persons that have or shall [destroy meadows] or over cut the forests."[55] The General Court in Plymouth limited the number and dates animals could graze on Billingsgate.[56] Yarmouth asked the legislature for restrictions on grazing on dunes and meadows because the destruction of the grass led to erosion. The colonial legislature then acted to prohibit grazing between December and March.[57] The prevention of stock grazing by fencing off land required an act of the General

Court, and by the second half of the eighteenth and early nineteenth centuries the General Court (and later the early state government) was receiving countless petitions from the Cape asking for restrictions on grazing on delicate dunes.[58]

By the end of the eighteenth century the Outer Cape experienced loss of forests and serious erosion because of deforestation and overgrazing of the dunes, yet there were still significant stands of woods elsewhere. South and West Sandwich, today's Bourne and Falmouth, were heavily wooded. Sandwich continued to export wood off Cape into the early nineteenth century, and in 1825 it was chosen for the site of the Sandwich Glass Works partially because its thick forests offered plenty of fuel for the kilns. There were several wooded areas in Barnstable, notably just south of the Moraine ridge, which ran from Sandwich to Dennis.[59] Even the Outer Cape had clusters of woods. Timothy Dwight noted on his travels across the Cape that there was a forest that began six miles north of Orleans and continued almost without intermission to Wellfleet.[60] The Human Society, a charity group that set up huts for stranded sailors along the coast, investigated the east coast of Cape Cod at the turn of the century; it found several areas along the coast where the forests were too thick for stranded sailors to find their way to homesteads. Several sections of the coast south of Provincetown were also found to be heavily wooded, and the Wellfleet coast was a "thick and perplexing forest."[61] A traveler to Provincetown commented in 1791 on the thick, pitch pine woods with trees higher than twenty feet on the south side of the settlement.[62] As late as 1807 Truro was described as having "lofty forests," although fifteen years later the town was importing wood from Sandwich. Nonetheless, eighteenth-century Cape Codders faced deforestation in several locations that drove up the price of cordwood and lumber dramatically. Without trees to hold the soil in land, erosion grew to be a serious problem. Within two generations the rich soils of southern Eastham had been depleted and were blowing away. Residents moved northward in search of better land.

FIRST MEASURES TO RESTORE DEPLETED WATERS AND LAND

Even the rich bounty of waters off Cape Cod faced problems that brought forth government action to temper the exploitation. Concern for shellfish led to extensive town laws limiting how much and when shellfish could be harvested—and by whom—in town waters.[63] Town and General Court regulations restricted when, where, and how fish could be caught both at sea and in inland waters.[64] In 1773 the town of Wellfleet restricted the catching of fish in the Herring River to only Mondays, Wednesdays, and Fridays.[65] Mill owners were required to open sluiceways to allow the free migration of fish to spawning grounds, and limits were placed on when they could be fished. In the middle of the eighteenth century, when the health of the important harvest of alewives began to worry town officials, agents were appointed in several Cape towns to make sure passageways to spawning grounds were cleared. During these times mills would have to cease operating, lift waterwheels, and open sluiceways for fish to pass through. Sandwich prohibited grinding from April 7 to May 20, while Barnstable instituted a thirty-day cessation of grinding and forbade anyone from taking alewives for a "foreign market."[66] Although these measures did not always succeed, and did stir up controversy, they reflect the willingness of these early Cape Codders to engage the issue of resource use.[67]

The forests, grasses, and waters of Cape Cod had so far provided its residents with their livelihoods, shelters, fuel, and food. Cape residents needed wood for building homes, barns, fencing, boats, and barrels, and for burning as fuel. They needed to feed their animals, and they needed the manure that well-fed livestock provided. For most of the Cape, although the demands for wood led to soil erosion, deforestation, and an increase in price, farmers were able, through careful husbanding of their woodlots, pastures, and plowed fields, to achieve, at least through most of the eighteenth century, a strained but sustainable relationship to the land. Although damages of overgrazing, overcutting, and maritime overharvesting were already evident, there were enough years of full abundant catches and good harvests to keep these worries at the back of Cape Codders' minds.

Despite the problems of erosion and deforestation the Cape greeted the nineteenth century with optimism. The Revolutionary War had disrupted shipping and fishing for the Cape, and even peace did not bring calm to conflicts at sea. Yet most Cape Codders looked forward to putting their amphibious world back in action. Orders for fishing boats piled up at boat- and shipworks. The Cape's whaling fleet that plied the North Atlantic, greatly diminished by the war, once again set sail for the whaling grounds that stretched north from Stellwagen shoals to the Arctic Circle. Cape fishermen sailed northeast for the rich cod-fishing banks. Initially these were small boats with crews of six to eight men who would fish the banks from April to September, but both crews and boats grew in size as the century ended. By 1790, despite having lost a number of vessels during the war, Provincetown sent eighty-eight men out in eleven fishing boats with a capacity of more than fifty tons each to bring home 8,200 quintals of cod.[68] Demand for salt increased, and in the 1770s the development of solar evaporation vats vastly increased the productive capacity of Cape salt makers. In 1789 Congress passed the first of several steep tariffs on imported salt. This encouraged Cape salt producers to dramatically expand production. By the beginning of the nineteenth century salt production dominated the shoreline of Cape towns. In East Dennis alone, salt vats covered more than three miles of coastline.[69] Merchants took their coasters out of Cape harbors loaded with whatever they could find that might command a price at an off-Cape market. On their way they made several stops, taking on and unloading different commodities before their eventual return home. Regular packet services developed in the last decades of the eighteenth century to simplify the process of moving people, mail, and goods between the Cape and a number of off-Cape locations.[70]

Although there were signs that Cape Codders had mined their environment to its limits, the Massachusetts Historical Society's 1790s descriptions of the Cape depicted comfortable and prosperous communities. Barnstable produced "good crops of Indian corn, not less it is supposed than 25 bushels to an acre on average and rye and other grain in proportion. Some of it is good for wheat and flax."[71] The town also produced "from 12 to 18 thousand bushels of [onions] . . . which are sold principally in Boston and the neighboring seaports."[72] These late-eighteenth-century commentators noted that such production of onions could deplete the soil, but hailed the presence of "their extensive salt meadows [which] enable them to keep large stocks in proportion to their pasture grounds . . . [where the] manure made by cattle fed on salt hay is much more fertilizing than that made from fresh."[73]

Historical collections include descriptions of Provincetown's "inhabitants in 1790 as employing twenty vessels in the cod fishery," and with "a great variety and abundance of excellent and profitable fishes," in and about its harbor.[74] Provincetown fishermen put out seines in the harbor and caught five to six thousand barrels of herring worth four dollars a barrel in the late fall, and another three hundred quintals of bass realizing four dollars a quintal. Over three hundred barrels of mackerel were pickled and sent on to Boston. Five Provincetown boats fished for lobsters that were sent off to New York or Boston. The herring was mostly sold locally to cod fishermen for bait.[75]

The Cape's late-eighteenth-century economy thrived on a combination of resources harvested from the land: grains and root vegetables; cattle and sheep,

Cape farming depended upon salt hay, which was brought from the marsh and used to feed cattle over the winter, allowing Cape Codders to maintain larger numbers of cattle than their pastures alone could sustain. Farmers then spread the manure on their fields to fertilize the sandy soil. Photo courtesy of the Wellfleet Historical Society.

dependent upon meadow grasses and marsh hay; fish, whales, oysters, and lobsters. The productive activities that supported those ventures included milling and lumbering; ship and boatbuilding, repairing, and outfitting; salt making; barrel making; barn, warehouse, and wharf building and repairing; and selling these resources to markets in the larger Atlantic world.

Many of these activities complemented each other. In combination they provided a comfortable, if dangerous, life for the Cape's growing population. Homes were extended, additions added, new farms opened up, new fields cleared. Wharves were lengthened and new ones built. Shipyards took on new workers and built ever-larger vessels. More trees were cut and more land was cleared and plowed. Although the 1643 Committee of Seven had reported that the Cape was just too far away to become the new center of the government for the Plymouth colony, the belief of its original members that the Cape would become a rich and prosperous place seemed, in 1790, to have been borne out.

4
MINING THE BOUNTY OF NATURE

By 1800 the American Revolution was over, and the preceding decades had established clear paths to prosperity and bounty upon which succeeding generations could build. So prepared, the Cape embarked on the new century with the optimism of a fisherman, tanked with ale, stepping on board for another run to Newfoundland's Grand Banks. No element in the Cape culture warned its residents that the consequences of their ways might visit them with destruction. The common law that secured property and formed the basis of their government told them that all that dwelt on the sea and land was property, whereas their religion told them they had a duty to subdue the earth (Gen. 1:28).

THREE LEGS OF PROSPERITY: Fish, Forests, and Soil

Timothy Dwight, the president of Yale from 1795 to 1817, saw this optimism when he reached the Cape in late September 1800. Sandwich surprised him with its prosperity. "The meadows are often brilliant," he wrote in *Travels in New England and New York*. "The arable land bears good crops of grains . . . among them wheat which not uncommonly yields well," as well as "good orchards," and along the shore a "considerable salty marsh yielding the inhabitants a large quantity of hay," and even forests that produced enough surplus wood that much of it is "carried to Boston."[1] (Dwight's posthumously published *Travels*, which included a good deal of socioeconomic observation concerning those regions, and especially the Cape, is perhaps his most renowned work.)

In Dwight's mind the Cape farmers had a difficult time with their "light and thin" soils, yet he found "neat and tidy" homes as he traveled east from Sandwich.[2] Barnstable was a prosperous community with comfortable homes, several of which "exhibit[ed] proofs of wealth and taste."[3] Near these tidy homes were "small barns in which lodged the salt hay destined to be the food of [the farmers' cows]"[4] In Eastham, Dwight was impressed with the productivity and prosperity of the inhabitants. Some of the land was "probably the best in the county yielding when well manured from 30 to 45 bushels of maze and 20 to 30 bushels of rye" an acre.[5]

The land produced grain, hay, vegetables, and apples, but Dwight claimed that the richness of the sea "tempted [Cape Codders] to seek for plenty and prosperity on the waves," and from the seas they "derived their wealth and much of their substances."[6] Daniel Webster, a Massachusetts senator in the years leading to the American Civil War and a regular visitor, noted of the Cape later in the century, "your labors are on the sea . . . your home is on the deep."[7] It was a sentiment that a Dennis resident shared when he noted, "Though we may boast of [a few] richly productive farms here surrounding the numerous pleasant dwellings, their owners generally farm the broader and more fertile Atlantic."[8] As one moved north from the elbow of the Cape at Orleans, "the principal concern of these people . . . lies with the ocean."[9] And just as the fertile fields and rich meadows sustained a simple but secure life for those working the land, the sea allowed other Cape Codders to live a "comfortable and thrifty" life as fishermen. Provincetown was the most successful of the fishing communities in Dwight's mind. "The cod fishery is pursued with great spirit and success. Just before we arrived, a schooner came in from the great banks with fifty-six thousand fish, about fifteen hundred quintals, taken in a single voyage."[10]

The Reverend Levi Whitman reported to the Massachusetts Historical Society in 1802 that Wellfleet had twenty-five vessels. Five were engaged in whaling in the Straits of Belle Isle and off Newfoundland. Of those, one vessel had a capacity to hold 100 tons, three had a capacity of 75 tons, and one had a 57-ton capacity. By comparison, the *Mayflower* was a 180-ton vessel. (In all cases the tonnage capacity refers to the number of casks of wine the ship could store in its hold.) Four vessels, each between 40 and 50 tons, engaged in cod and mackerel fishing. Four 30-ton vessels were employed in transporting oysters to Boston and other ports. Twelve smaller vessels with a capacity of 12 to 16 tons engaged in fishing around the Cape.[11]

By 1834 Provincetown fishermen were bringing in annually forty-five thousand quintals of cod and another seventeen thousand barrels of mackerel. The Outer Cape towns were not the only ones harvesting the bounty of the seas. In 1837 Dennis had eighteen vessels bringing in annually over nine thousand quintals of cod and nearly five thousand barrels of mackerel.[12] Yarmouth, at the beginning of the century, had six wharves on the Bass River, home to one brig that sailed to the West Indies and ten coasters of thirty to forty tons that sailed north to Boston and south to Connecticut, New York, and the southern states as well as to the West Indies. Yarmouth also had ten vessels fishing the Straits of Belle Isle, the Shoals of Nova Scotia, and Nantucket Shoals.[13]

"Upon the whole," Dwight found, "this unpromising [Cape] sustains more inhabitants and furnishes them with more comfortable means of subsistence than a stranger would be easily induced to imagine."[14] And not only farming and fishing contributed to the Cape's prosperity, but also salt making. A member of the Massachusetts Historical Society in 1802 was impressed with Dennis, calling it a "flourishing place" that prospered because of good farmland and successful fishermen, but principally because of its saltworks.[15]

The Cape that Dwight visited in 1800 was on the verge of what its residents came to feel was its golden age. More and more neat, thrifty, and comfortable homes were built over the next half century. And those who found new wealth on the land and sea had constructed even larger and more spacious homes. In reflecting on the Cape in 1839, the venerable historian and Cape native John Palfrey said that the Cape "meets our view with all the tokens of being the seat of an intelligent, virtuous, efficient population. We see its harbor[s] a scene of cheerful activity. In the fields we look at substantial harvests."[16]

The Cape's prosperous golden age rested upon three legs: soil, forests, and fish. Each depended upon the outcome of a contest between evermore encompassing commerce and the sustainability of natural resources after harvesting. During the half century of the golden age, Cape Codders expanded the reach of their economy. When signs of resource exhaustion appeared, they regarded them as something they could manage.

OCEANS OF "INEXHAUSTIBLE FERTILITY"

Cape whalers had been searching for leviathans since the end of the seventeenth century. As the industry evolved in the eighteenth century, whales were so frequent in Cape waters that Harwich whalers went after them in small sloops. But the whales that had been common in the bay and on the shoals just to the north of Provincetown grew less common into the nineteenth century.[17] Captain Atkins, a retired Wellfleet whaler who remembered bay whaling in his youth, claimed that in the 1720s whalers were taking seventy to one hundred whales in the bay every season and bringing them to shore for trying (i.e., boiling down the blubber).[18] The slaughter of the bay whales would put an end to this type of whaling. Reverend Whitman reported to the Massachusetts Historical Society in 1802 that in the early decades of the eighteenth century more than two hundred Barnstable men, mostly farmers, worked the "whale bay fishery" after the fall harvest, "but few whales now come into the bay and this kind of fishery has for a long time by this town at least been given up."[19]

By the end of the eighteenth century small sloop whalers were out of business, and only larger vessels running out of deeper harbors, which could manage trips of longer distance and time, remained in whaling.[20] In 1802 Reverend Whitman also informed the historical society that whalers out of Provincetown and Wellfleet searched for their catch in the Strait of Belle Isle between Labrador and the northern tip of Newfoundland because no whales were available in the bay.[21] Even this long haul provided no guarantee of a catch. Fearing that they would be unable to find whales in the region that had been whaling grounds since the early sixteenth century, the whalers loaded their ships with salt, hooks, and line so that "what they want in loading with oil, they can make up in cod fish."[22] But these whaling vessels did not set out with the purpose of catching cod. Whaling was far more lucrative. During the first half of the nineteenth century the demand and price for whale oil grew dramatically. Huge fortunes were to be made for those who captained or owned a share in a whaling boat. In 1843 a single whale yielded ten thousand dollars in oil and bone.[23] And yet whalers continued to load the salt, hooks, and line because the supply of North Atlantic whales was diminishing.[24]

Whaling ships had to travel farther and farther from port and stay at sea far longer to realize profits. Sighting a whale from shore, sailing out for the kill, and hauling it ashore for trying was an experience of the grandfathers of nineteenth-century whalers. To bring home whale oil in the nineteenth century was a heavily capitalized operation. The ships needed to be provisioned for three or four years at sea. And the ships needed to be large and powerfully built to withstand the buffeting seas—especially in the South Atlantic—and needed to be supplied with plenty of spare equipment.[25] Heavy storms that swept through the seas these whalers traversed often ripped sails and rigging to shreds. Spare sails and rigging were essential. Whalers would need to set down a small whaling boat from the mother ship to chase the whale, and then thrust a harpoon with a rope attached into the whale's back. The whale then took the whaling skiff on the "Nantucket sleigh ride," until the whale tired. The harpooner would then thrust a deep and fatal spear into the whale's lung or heart. In this process the skiffs were often badly damaged or destroyed. At sea and far from a friendly harbor, these boats required constant repair onboard ship. This task involved not only a carpenter but also a need for a hefty supply of lumber.

Wellfleet had a whaling fleet of thirty vessels at the end of the eighteenth century but saw fewer and fewer whalers leave its harbor each year.[26] The demand for any whale oil was high, but sperm whale oil commanded the highest price. Sperm whales could be found across the oceans of the world, but they were most prevalent in the South Pacific. Just getting to the whaling grounds required months of sailing through treacherous seas. Although Cape whalers joined in this wild and dangerous pursuit of the great leviathans, it was the better-capitalized vessels leaving Nantucket and, especially, New Bedford that dominated this lucrative business in the nineteenth century.

New Wealth in New Waters

Cape fishermen found other wealth at sea, however, and new waters in which to find it. Across the Cape, fishermen left ports for the rich cod and haddock fishing grounds of the Grand Banks, located southeast of Newfoundland, while other boats followed the trails of mackerel as they moved up and down the coast from Georgia to Maine following schools of menhaden.[27] Closer to Cape fishermen's ports was Georges Bank covering ten thousand square miles off the coast of Nantucket. Cape Codders knew there were fish on the bank, but catching them proved difficult.

To fish for cod in the early nineteenth century a sloop or schooner drifted while fishing lines with dozens of baited hooks were lowered over the side. The currents and winds crossing the Georges Bank were far more treacherous than those on the banks further north, especially for ships drifting about hunting for fish. Georges Bank had shifting shoals so shallow that breakers would crash over them in storms and ships would run aground. Until the 1830s, when a Gloucester crew successfully fished Georges Bank, American fishermen avoided it because of its violent currents, shoals, and massive sudden storms. The haul of that Gloucester boat convinced Cape fishers it was worth the risk to venture out to these turbulent waters, and yet the risk often proved deadly. On October 2, 1841, a huge Northeaster quickly formed and swept down on the Cape fishing fleet on Georges Bank. Realizing a storm was moving, the fishermen attempted to run to shore, but the winds and currents that plagued Georges Bank prevented their escape. The winds tore apart sails and riggings and drove the fleet on to the Nantucket Shoals. Among those lost in that storm were fifty-seven Truro fishermen.[28]

Despite the dangers, Cape fishermen continued to sail to the rich fishing grounds of the various banks off their coast for the cod and haddock. Unfortunately, the traditional Cape fishing schooners did not prove sufficiently seaworthy on the treacherous Georges Bank. With their rounded bows, broad sterns, and short spars, they were too slow to get off the bank when the weather turned foul. In response, boat builders in Provincetown and then across the Cape began building a leaner Georges Bank schooner with long spars, tapered sterns, and sharp bows. Even with the new schooners, fishermen on Georges Bank courted death. Yet the catch made it worth the risk. Cod and haddock salted and dried well. Soaking the dried salted cod in water and then cooking it up into a brothy fish stew created an easy-to-prepare, tasty, high-protein meal. Planters in the West Indies and the Carolinas looked to salted cod for a cheap and nutritious food supply for their slaves. City workers and sailors across the globe found salted cod a nutritious alternative to beef jerky. For Catholics away from a fresh supply of fish, cod stew met the Church's requirements for Friday meals.[29]

By the 1850s fishing shifted to large two-masted schooners with crews of a couple of dozen men and several dories on board. Upon reaching the fishing grounds the mother ship lowered the small one- or two-person fishing dories to spread out and cover a large fishing space. Fishermen would then lower lines overboard much as they had off the schooners themselves. The dories reduced the concentration of fishing lines in one limited space and allowed fishermen to cover a much larger area. By the late 1850s Cape fishermen adapted the French technique of long-lining.[30] The dories of long liners carried dozens of

Until the 1930s, sailing schooners, such as this one in the harbor, made up the Cape's fishing fleet. Once on the fishing banks, the boat's crew would lower fishing dories that would spread out around the schooner. Photo courtesy of the Library of Congress.

baited hooks strung out on lines. The dory men would lower them overboard. After a wait, they retrieved the lines and hooked fish into the dory. They restrung the hooks with bait to repeat the process. Once the dory was full of fish the dory men would row back to the schooner and unload the fish, which they would split, disembowel, and salt. These partially preserved fish were brought back to the home ports, where the fishermen would throw them into shallow water on the flats of the beach, wash them, and cover them again with salt. Then they laid them out to dry on flakes, frames that stretched out behind the shoreline, from Falmouth to Provincetown. The flakes were usually branch and bush twig platforms three to four feet wide, twenty to fifty feet long, and built two feet above the ground. The cod dried as it rested on the flakes, with the sun shining directly down on them and reflecting its heat up from the sea and sand.

Bait Fish, Food Fish, or Both

Good prices for catch and new places to fish encouraged Cape fishermen to adapt the new long-line dory techniques. More fishing and the newer

Dory fishermen lower long lines of baited hooks over the side, to be hauled in by hand. Photo courtesy of the Library of Congress.

techniques also meant fishermen needed more bait. The primary food fish caught by Cape fishermen were mackerel, haddock, and cod. Herring was an important food fish as well, but it was also used as bait for other fish, which Cape fishermen needed.[31] The supply of bait for cod fishermen was usually made up of a variety of herring (including menhaden, alewives, and river herring) as well as squid or large clams.[32] This bait came from the near-shore or from rivers. Each morning in the summer small catboats would leave the shallow harbors of Cape towns, and fishermen would attempt to gather up menhaden either in their seines or pound nets, or by snagging them on hooks and line. Thoreau noted that in July "schools of menhaden . . . rippled the surface," with the sea "spotted with them far and wide such is its inexhaustible fertility."[33]

In the late fall and winter months, Cape Codders dug for clams. Clam bait was in such demand that by the late eighteenth century Eastham was trying to reserve them for its own residents. In 1785 the town outlawed trespassing by nonresidents on its flats for gathering clams for bait.[34] Despite real local shortages of clams in the eighteenth century, many Cape Codders continued to believe clams to be "inexhaustible."[35]

Anadromous Fish

In the spring Cape Codders found a source of bait readily at hand when the rivers and streams ran so full of fish that one could scoop them up with a pitchfork. When food supplies were low after the long winter, the river herring (*Alosa aestivalis*), alewives (*Alosa pseudoharengus*), and shad (*Alosa sapidissima*) that migrated up the streams and rivers of New England had been crucial for the diet of Native Americans. After the first few winters in the New World, the Pilgrims also came to depend upon the annual upstream migrations of schools of fish in the thousands; they were easy to catch and provided an im- portant source of protein for poor families as well as food for their pigs.[36] When fishermen began catching cod, they harvested these anadromous fish for bait. So important were they for bait that when Wellfleet, in 1772, limited the catching of fish from the Herring River to Mondays, Wednesdays, and Fridays it made fishing for "codfish bait" the exception,[37]

Herring, alewives, and shad were so easy to catch because they had to negotiate the narrow rivers and streams on their way to spawning ponds. Early colonial settlers recognized, not only the importance of these fish as a source of food, but also how vulnerable they would be if access to the spawning grounds were cut off. Owners of milldams were specifically instructed to guarantee that the waterway would be opened for fish migration.[38] Well into the first two decades of the nineteenth century, local towns and the state government alike aggressively sought to protect access to spawning grounds and limit the harvest of these important fish.[39] Cape Codders not only petitioned the legislature for acts to protect these fish migrations, but they also dug channels connecting inland ponds to the sea in hopes of increasing the spawning areas.[40]

Mackerel Barrels, Oyster Beds, and Eider Down

Cod was the primary commercial food fish for Cape Cod fishermen, but it was not the only food fish sought out. The other major fish that occupied the attention of fishermen was mackerel. Mackerel were oilier and not as easy to salt, dry, and store, but could be pickled in brine in airtight barrels. Although their schools covered huge distances as they followed their prey fish north and out to sea, mackerel also tended to school closer to shore and offered fishermen with only a catboat or dory the opportunity to catch and bring to market large numbers of fish. Mackerel were either sold fresh locally or sent to distant markets pickled and barreled. As Cape fishermen brought home fish to be packed for off-Cape markets they needed barrels—and hundreds of them. Along with boat works, rope works, and sail works, the Cape's fishing ports also supported cooperages that often used local wood.[41]

Meanwhile, the oyster beds dead in the late eighteenth century, due to a combination of overharvesting and disease, were rebuilt by bringing up oyster seeds from the Chesapeake and planting the seeds in the bays and harbors of the Cape, especially in Wellfleet Harbor.[42] This was an expensive operation and eventually involved actually farming the oysters.[43] But the hundreds of thousands of seeds that were brought north returned hundreds of dollars of profit for those able to finance this new aquaculture. These were not oysters to be restricted for the benefit of local families, as the 1773 ordinance proclaimed. These were oysters raised for the Atlantic market. The world-famous Wellfleet oysters that Queen Victoria had shipped to her in London were no longer of Wellfleet origin, but they were Wellfleet-raised, and they produced a handsome profit for their Wellfleet producers.

Not only did Cape Cod sailors find new grounds to fish, they also found new products in those fishing grounds. While fishing in the Strait of Belle Isle (between the Newfoundland mainland attached to Quebec and the island of Newfoundland), Cape fishermen began killing Eider ducks that flourished along the Labrador coast. The need for comforters to protect sleepers from cold New England winters created a strong market and a good price in Boston for the feathers of these ducks. Cape Codders brought hundreds of thousands of bags of feathers into Boston and many nineteenth-century ship captains made their fortune from the duck feathers.[44]

BUSTLING HARBORS AND THE BUSINESS OF TRANSPORT

The wealth of the sea also translated into expanding and developed harbors. Warehouses, stores, and rigging and blacksmith shops joined the wharves,

Before the tourist economy took root on the Cape, the shoreline was a messy workplace. Here pound net fishermen pull in fish from weir nets they had placed along the shore. Today such shorefront property is far too expensive to be used for such activities. Photo courtesy of the Library of Congress.

ship and boat works, and barrel and sail making facilities in towns across the Cape.[45] An early-nineteenth-century observer noted that "at the head of each of the landing-places, a store and all around warehouses, pyramids of new barrels, workmen, and all having a look of most especial thrift and contentment."[46] Theodore Brown's Wellfleet wharf, where he repaired and built small and medium-size boats as well as loaded and unloaded goods, was overshadowed in 1835 when investors built Commercial Wharf. Central Wharf in Duck Creek and Enterprise Wharf in Chipman's Cove followed, also built that same year. These larger wharves accommodated the increased seagoing traffic and the expanding fishing trade.[47] By the early nineteenth century Yarmouth had six wharves on the Bass River that held a store for marine goods as well as blacksmiths' shops, ships' carpenters, and sail makers.[48]

During the eighteenth century people had traveled to and from the Cape and between Cape towns as passengers on boats carrying other goods. The Reverend Mellen reported to the Massachusetts Historical Society in 1794 that of the thirty-three vessels in the town of Barnstable most were schooners engaged in fishing, but that two or three of these vessels engaged in the "merchant service."⁴⁹ A person heading to Boston could go to the harbor and sign up to be a passenger on a Boston-bound boat that had already been loaded with fish, wood, or grain. By the end of the eighteenth century, Cape ship captains had developed regular packet service between the mainland and the various towns and villages of Cape Cod. These packet services served as the main conduit between the Cape and the expanding economy of New England.

Initially, when a packet arrived at one of the towns a packet service flag would be flown from a high point, announcing the arrival of the vessel from Boston. The day before the packet was scheduled to depart, another signal would be hoisted so that those with goods destined to the Boston market could load them aboard.⁵⁰ By the early nineteenth century weekly packets from Boston served most Cape towns, and some towns had two or three packets working per week. As the century progressed, they became more regular and frequent. Packets ran from the south shore to New Bedford and Providence. Packets also sailed to Martha's Vineyard and Nantucket. By the middle of the nineteenth century the packet schooners were regularly carrying between twenty-five and fifty people crowded into cabins and on deck. These boats also transported barrels and boxes of food and goods as well as the mail.⁵¹ A one-way trip to Boston, which could take between six hours and two days depending on the weather, cost the traveler roughly seventy-five cents.⁵²

The focal point of life on the Cape shifted to the harbors in the nineteenth century. As the Cape's economy became more intertwined with the wider commercial world, more and more of its residents were doing more than fishing or farming. Marion Crowell Ryder's great-grandfather, Levi Crowell, began in the coasting trade in 1827, running a locally built schooner between Yarmouth and Boston, New York, and Portland. After his marriage, he built a store beside his Yarmouth house, stocked with goods from his schooner runs and overseen by his wife. The store prospered so that Crowell gave up sailing schooners to focus on the store and invest in a fleet that fished the Grand Banks for cod. In 1832 he built a wharf on the Bass River and a three-story warehouse. He used the lower two stories for goods he brought by boat for the store, especially goods for outfitting his fishing fleet. He leased the top floor to the Southwood and Light Sail Company, which made and repaired sails.⁵³ In 1849 Stillman Kelley, who had previously fished out of Harwich, moved to Dennis to open "a store and engage in the fishing trade." Within a year he opened a general store in partnership with Seth Sears on the wharf and purchased six fishing vessels. He took his fish to outside markets and brought in goods for his store on the wharf. In 1851, reflecting the growing expense and shortage of cordwood on the Cape, Kelley and Sears added coal to their stock, and in 1852 the company established a lumber yard on the wharf, bringing in the lumber by boat. That same year they took on Nathan Sears as a partner and had seven more boats built to add to their coasting and fishing fleet. In 1857, following the death of Seth Sears, Kelley and Nathan Sears bought a store in the village center and transferred their small goods to that store. When the two remaining partners finally sold the company in the second half of the nineteenth century, they were totally out of the fishing business and focused entirely on transporting coal, lumber, and general merchandise.[54]

Shipbuilding

Cape carpenters had long been turning their skills to boatbuilding. Cape builders constructed boats to meet the needs of their expanding amphibious culture. In the late eighteenth century, Jesse Crosby began using local wood to build boats for near-shore fishermen. Nantucket Sound waters were rich with fish—not cod, but mackerel and especially herring and squid, which were in high demand as bait for those fishing for cod.[55] Nantucket Sound waters were shallow, as were the harbors on the south shore of the Cape. Crosby began building boats in Centerville and then moved on to Osterville, where he developed the Crosby catboat. These shallow-drafting, wide-beamed, gaff-rigged boats with masts well forward were particularly suited to Cape waters. Although they were not fast, they were exceptionally stable, held a significant load for their length, were easily sailed solo or with a small crew, and could be built small or large. Local oak, pine, and cedar were the main building materials for these boats. By the early nineteenth century an observer noted "little fleets constantly visible in the distance following the schools of their prey."[56]

The Crosby boatyards were not the only ones producing Cape vessels. Yards from Falmouth and Sandwich to Truro and Provincetown were hauling in wood and building vessels ranging from the small catboats to large two- or three-masted schooners. Reflecting back upon the first half of the nineteenth century, one observer noted that the building of coasting craft (the boats that moved the products of the Cape to markets), was a "staple industry" for Cape towns.[57]

Sandwich produced a 400-ton ship worth $15,000 in 1837. In 1855 Barnstable was still producing fifteen small boats a year, as was Chatham. Harwich produced forty and Provincetown produced seventy. Even Eastham, a town lacking a significant harbor and thus with a smaller proportion of its population dependent upon the seas, built a 101-ton schooner in 1812 using timber "cut upon the surrounding territory."[58] Wellfleet in 1800 built the 100-ton schooner *Free Mason* out of local wood including black locust, which had been introduced to the Cape by colonial settlers. The *Free Mason* was built on Bound Brook and pulled out to sea during exceptionally high tides.[59]

Although boat builders worked in most Cape towns, Dennis was the center for the Cape's nineteenth-century boat production. Asa Shiverick's ship works not only turned out major schooners for local fishermen but also constructed the huge new and fast 1,000-ton clipper ships that were setting speed records for voyages to Asia, Africa, and Europe. The Shivericks were not the only large Cape builders turning out these huge new ships, but the eight they produced were the most renowned. These new clipper ships consumed massive amounts and various kinds of wood—far more and of higher quality than was available on Cape by the middle of the nineteenth century. Furthermore, they were not the only boat builders calling for more and better quality lumber than could be supplied by local forests. As early as 1825 the records for the dock in Yarmouth indicate large amounts of ash planks being imported from Maine for the Yarmouth boatworks.[60] Even the smaller boat works found that less and less local wood was available.[61]

The small catboats as well as the large schooners needed not only wood for their construction but rigging, tackle, gear, and sails.[62] The making of sails was a major port activity and required significant skill as well as space. The suits of sails were either drafted from blueprints or copied from old sails on the floor of the lofts of boat-yards or warehouses.[63] Initially Cape sail makers used sailcloth produced at mills in Marston Mills or Sandwich. By the nineteenth century demand for sailcloth vastly outstripped the capacity of these mills, working as they did off the limited power of the Cape's small streams. As a result, bolts of sailcloth were imported from the mainland. Besides sail-works, the rear of most Cape wharves had extensive rope works. Once built and rigged, except for the coasters, these boats, whether catboats or schooners, were primarily about catching fish, which in this age before refrigeration needed to be preserved in salt.[64]

Salt

Salt production through solar evaporation, initially developed at the end of the eighteenth century by the Sears family of Dennis, expanded greatly in the nineteenth. Seventeenth- and eighteenth-century fishing and farming families produced salt in small inexpensive operations involving large kettles and cheap or free cordwood. Salt water would be hauled to the kettle, usually located along the water line; cheap pinewood, which burns fast but produces a hot flame, would be fired under the kettle and the water boiled off to leave salt crystals that would be scraped out before filling up a new load of water. The process was slow but not capital intensive as long as pine cordwood was plentiful. The salt it produced may have met the demands of family- and small-scale fishing, but this method could not meet the dramatic increase in the demand for salt that accompanied the expansion of fishing. By the nineteenth century, cheap or free cordwood was no longer easily available, especially in Cape towns east of Barnstable. To meet this growing demand, fishermen increasingly turned to imported salt. At the end of the eighteenth century this need led to the production of large Cape saltworks. When Timothy Dwight visited the Cape in 1800 he found forty-seven saltworks in Dennis, fourteen in Barnstable, and four each in Yarmouth, Sandwich, and Falmouth. Harwich had twenty working saltworks, whereas Orleans had eleven, Eastham twelve, Chatham six, Wellfleet two, Truro one, and Provincetown ten. These operations employed thousands of workers and, in 1802, produced over forty-one thousand dollars' worth of salt.[65] Wendell Davis informed the Massachusetts Historical Society that by 1802 over one hundred thirty thousand dollars was invested in salt making, producing a return on investment of more than 30 percent.[66]

The combination of three factors—the solar evaporation innovations of Captain John Sears, a national government tariff along with a bounty for fishermen to help offset the added cost of the tariff, and the soaring demand for salt—encouraged the rapid building of saltworks along the shores from Sandwich to Provincetown. These saltworks were not simple backyard operations. Windmills pumped seawater into the water room, which was a vat eighteen feet by thirty-six feet, ten to twelve inches deep, raised two to four feet above the ground, and built of pine one-and-a-half-inches thick. After significant evaporation taking three or four days, the water was then pumped into the pickle room, a vat twenty by twenty feet, for further evaporation and where lime and other minerals would deposit. This highly salt-concentrated slurry was pumped into the twenty-by-twenty-foot salt-room vat where crystals would form. Finally these crystals would be raked out and taken to the salt room where the last of the moisture evaporated and salt crystals were harvested. The total process required three weeks. As each vat was emptied of its contents new water was pumped into the vat. It took 350 gallons of salt water to make a bushel of salt.

Over the vats were low retractable wooden roofs that were pulled back when the sun was shining. In the spring the pickle vats were cleaned of the slurry at the bottom, which was boiled down to Glauber's (sodium sulfate) and Epson salts (magnesium sulfate).[67] These were expensive operations taking up large tracts of shorefront land.[68] In 1800 the cost of building a saltworks using local lumber was $1.00 to $1.25 a vat foot, measuring one foot by ten feet. Even with the use of local wood, the material for setting up a saltworks consumed two-thirds of the total cost while labor represented one-third.[69] In 1810 Dennis alone had six hundred fifty thousand feet of vat space.[70] Even the one-and-a-half-foot pine boards used for building the vats were expensive, and because the vats were built at the edge of the shore, they often had to be rebuilt after severe winter Northeasters or the less frequent but just as damaging late summer-fall hurricanes. On September 23, 1815, a huge hurricane swept on to the Cape, and, unlike earlier hurricanes, this one did substantial damage to the infrastructure that had been built up since the end of the eighteenth century. "Salt-works [were particularly] destroyed; . . . nearly all on Buzzard's Bay—and they were numerous—were swept away."[71]

Saltworks continued to be constructed across the Cape throughout the early decades of the nineteenth century. By the 1830s there were more than 442 salt operations on the Cape producing over a half-million bushels of salt annually and realizing a profit of approximately two million dollars.[72] This was a handsome return on investment, but it was not a return within the reach of the small farmers or fishermen's families who previously harvested salt from their backyard kettles. A new saltworks operation on Billingsgate Island in Wellfleet was incorporated in 1821 with fifty thousand dollars in capital.[73] The solar evaporation systems took less of a toll on the region's forests, but the capital costs were well beyond the means of most Cape Codders. Much of the capital for these works came from profits accumulated by the most successful local whalers or merchants.

Truro, which had one saltworks in 1802, had thirty-nine producing almost 75,000 bushels of salt per year in 1837.[74] In 1845, Barnstable had twenty-four saltworks producing 21,000 bushels, Falmouth had forty-two works producing 24,000 bushels, Orleans had forty-six plants producing 17,000 bushels, and Provincetown had seventy-eight saltworks producing 26,000 bushels. Yarmouth had sixty-five plants, which produced 74,065 bushels. Dennis had eighty-five saltworks, which turned out 34,000 bushels of salt yearly; Eastham's

Mining the Bounty of Nature

Saltworks, which covered much of the shoreline from Provincetown to Sandwich, required a great deal of wood and were in constant need of repair. Increased prices for wood and competition from supplies in Europe eventually drove the saltworks from the Cape shores. Photo courtesy of the Cape Cod National Seashore.

thirty-five plants produced 17,320 bushels of salt and Chatham had fifty-four works, which produced 18,000 bushels of salt. These were not little operations. They accounted for over two million dollars in investment by midcentury. But the pinewood needed to build and maintain these saltworks increasingly had to be imported from Maine because the eastern Cape towns had cut down much of their forestland.[75]

THE NEW FARMING

As Cape Codders searched for new ways to fish and for new fishing grounds, their neighbors found new ways to turn a profit from the land. In the eighteenth century most Cape homes had significant gardens and often their own cows, pigs, and chickens. But by the nineteenth century, the owners and craftsmen who worked along the wharves and harbor centers were too busy to be part-time farmers and part-time boat- and shipwrights, clerks, blacksmiths, coopers, and sail makers. For their part farmers increasingly found markets for their turnips, milk, onions, fruits, and eggs in the villages that clustered around the harbors. At the same time the markets for meat, vegetables, fruits, cereals, and fish grew in the commercial port cities of the Northeast. Cape Cod farmers brought more and more surpluses to their town harbors to be consumed there or shipped off to wider markets. In the early nineteenth century Eastham was exporting over three thousand bushels of wheat as well as onions and turnips and cranberries. Brewster, with some of the most fertile Cape soil, was producing a surplus of fruits, cereals, and vegetables and sending it to the mainland. Eastham and Brewster were not alone; almost all the Cape towns were shipping bushels of onions and turnips as well as fish and oysters off-Cape.[76] And until the second half of the nineteenth century they sent them off-Cape by boat.

By the early nineteenth century much of the forests of the Outer Cape were heavily cut and cordwood was in short supply.[77] By 1809 a cord of wood sold for seven dollars, a significant sum for those with little extra cash.[78] But the Cape did have an abundance of peat bogs. Dried peat burned well enough and was available, although it required some hard labor to extract it from the bogs. Most Cape farmers had access to peat, which had formed in the acidic low-oxygen environment of the bogs littering the Cape landscape.[79] Peat was as cheap as the plentiful labor necessary to extract it.[80] In the early nineteenth century, drying peat became as common a sight behind Cape cottages as stacked cordwood had been in the eighteenth century.[81] But the Cape's substitution of peat for cordwood was short-lived.[82] Another competing demand arose for the bogs other than peat, and it was far more remunerative.

Cape Codders, both Native Americans and white settlers, harvested wild cranberries in the shallow kettle holes of the Cape. Although prolific by the standards of wild berries, a wild bog did not produce many more berries than would meet the needs of a few families. In the early nineteenth century, a Dennis farmer discovered that transplanting wild cranberry vines into bogs and covering the vines with sand could vastly increase their productivity. Although most Cape farms had some bog land, converting that bog land into profitable cranberry operations was expensive and beyond the reach of most Cape farmers. In the 1830s it took roughly one thousand dollars to clear and plant a cranberry bog. Drainage levees as well as pumphouses, windmills, floodgates, and ditches needed to be built, and it took three years before the berries could be harvested.[83] But if one had the capital to do all this, the rewards were generous. If the market, weather, and lack of pests were favorable, a sixteen-acre improved bog yielded eight thousand dollars in cranberries, which could mean returns of up to 100 percent.[84]

Cape farmers had always raised sheep, but in the nineteenth century, with the increased price of wool, the number of sheep grazing on the dunes and meadows of the Cape, particularly the Outer Cape, grew dramatically. In fields from Falmouth to Truro, grains (wheat, rye, and oats, which were more cheaply raised in the western United States) increasingly gave way to onions, turnips, and flax.[85] In the early nineteenth century Cape farmhouses began replacing their fireplaces with the more efficient cast iron wood-burning stoves. These stoves reduced the amount of cordwood necessary for a home from thirty cords to ten to fifteen cords per year. This reduced the necessary size of a wood lot from thirty to forty acres to ten to fifteen acres. Farmers converted these extra acres of woodlot to pastures and cropland, a practice that initially brought in more income to the farmer but also opened up more land to erosion.

MILLS AND MANUFACTURING

The expansion of the nineteenth-century economy brought changes not only to the harbors and farms of the Cape, but to its streams as well. Increasingly, millers turned from grinding grain to capturing waterpower for other more lucrative activities. Sandwich, which had housed gristmills since its earliest settlement, built two mills in the early nineteenth century for cotton and sailcloth.[86] Cotton and sailcloth manufacturers had different concerns than grist millers. They were not working with local farmers and taking a toll from the grain milled but, instead, focused on larger markets. Shutting down to let fish migrate up to spawning grounds seemed an antiquated idea to these investors in manufacturing. Increased production for the expanding economy was their focus. It was a focus with which many, but not all, Cape Codders concurred. In the early nineteenth century Ephraim Ellis of Sandwich, who dammed a river that ran through his property to power his nail factory, objected that the town forced him to open his dam to allow fish migration. He eventually took his complaint to the General Court, arguing that since the river ran through his property, the town did not have a right to regulate what he did with it. On this occasion the General Court backed the town's traditional riparian rights, but the issue was clearly engaged.[87] Frederick Freeman, looking back from 1859 over the preceding fifty years, argued that in the colonial and early national period "it was the object to encourage the access of this fish to the ponds they have been accustomed to frequent, but the only stream [in Sandwich] of which they have held possession in any considerable numbers to the present time [1860] was the Herring river," because "their natural course having incautiously been obstructed by dams and mills."[88] Sandwich was not the only town facing increased pressure to dam its streams and build mills, whether for weaving cloth for sails, cottons, and woolens or for powering nailworks and ironworks. Marston Mills, a village within Barnstable, gained its name because of its mills, which were increasingly engaged in activities other than grinding grain for local farmers.

If protecting herring runs seemed less important to those who were building mills for cottons, woolens, and sail cloth, others cared deeply about it, particularly those who sold river herring and alewives for bait to cod and mackerel fishers or those who looked to the spring fish runs to add protein to their own or their livestock's diet. Early nineteenth-century towns of the Upper Cape faced conflict between those supporting common-law rights to the fish and those who saw the future in laissez-faire economic expansion. In 1806 a new mill was built in Falmouth for spinning woolen yarn.[89] The new milldam built on the Coonamessett River blocked the migration of fish into Coonamessett Pond. Those who fished the river for bait and food demanded the milldam be lowered to allow fish to migrate up the pond. Other town residents supported the mill owner in his refusal to lower his dam for the fish. The pro-fish group became known as the Herring Party; they were mostly the town's poorer residents who saw the fish migration as a common-rights issue.[90] The Mill-dam Party, those supporting the milldam, rallied to defend the dam against the town's poorer folk, arguing that the Herring Party was made up of the poor, who "contribute but a small part of the burden [of the town] while the mills are really a great convenience and advantage to the town," more than "could be benefited by fish passing in said stream."[91] Taking actions to defend their common rights, the Herring Party hauled out a cannon and attempted to blow up the dam with cannon fire. Unfortunately for the Herring Party, the cannon exploded and killed the gunner. His death deflated much of the enthusiasm of the Herring Party, and the Mill-dam Party managed to protect the mill despite the blockage of the fish migration.

The defeat of the Herring Party in the standoff at Coonamessett Pond did not end the controversy over milldams blocking fish migrations.[92] However, as the century progressed, support for common-law rights faded in face of the expansionist economy.[93] Although herring continued to migrate up Cape streams, that migration became more and more difficult with larger dams and later the dikes that were built to facilitate railroad lines. Already by the middle of the century, Henry David Thoreau noted that "Herring Rivers" on the Cape would soon "be more numerous than herrings."[94]

The economic integration of the Cape into the growing national and international economy not only encouraged Cape Codders to mine resources both on land and at sea with greater intensity, but also (although others found this trend threatening), to look to new organizational forms to accomplish their purposes. In the 1830s, private individuals looked to the state to incorporate companies to take control over inland fisheries. In 1837, Rich Sparrow petitioned the legislature for the right to establish a corporation that would oversee private development of inland fisheries on the Cape. Although some Cape Codders felt that Sparrow's incorporation would cut off their access to resources, and thus fought his attempt, it found favor with the General Court.[95]

Fisheries were not the only area where private corporations were being used to bring greater control over the resources of the natural world. Corporations were formed to build large wharves and to dike salt marshes to convert the land to sweet or fresh meadowland. Despite the importance of the salt hay available in Barnstable's Great Meadow for the town's small farmers, the state incorporated the "Great Marsh Diking, Water Power and Fishing Company" in 1850 "to reclaim for more useful purposes the vast amount of salt marshes that indent the bounds of the Cape." The company proposed to construct a dike from Calves Pasture to Sandy Neck "to prevent the flow of salt waters . . . for the purpose of draining the marshes there situate and converting the same into meadow or tillage land."[96] Access to salt marshes had, since the time of first settlement, been vital to residents for gathering salt hay for animal fodder, but fresh meadowland was more valuable than the already-prized salt marsh hay. Since diking streams involved common law, by chartering the new corporation the General Court accepted the right of the corporation to infringe upon traditional rights. Although the General Court was sympathetic to granting rights to dike and incorporating companies to do so, nature proved more difficult to maneuver than the legislature.[97] The several attempts made in the first half of the nineteenth century to dike off salt marshes to convert marshland to active productive farmland met with limited success. So, it was the coming of the more heavily capitalized railroad in the second half of the nineteenth century that did the most damage to existing salt marshes.

If diking marshland was not a success for Cape Codders, for many it seemed the only thing that was not succeeding. Fishing had always been dangerous and carried the risk of the large schools moving to new locations, but for the most part, those who took to the sea for fish and whales found them, albeit further away. With their catch, fishermen found prosperity and a degree of comfort. As late as 1855, whaling and fishing products accounted for over three million dollars and represented 60 percent of the region's income.[98] Many a lad of fourteen years, like Lorenzo Dow Baker, began his life "before the mast" and became the captain of his own vessel by the end of his twenties. The expansion of fishing in the fourth decade of the century meant a growth of boat and sail making. The expanding commercial activity to distribute products of the land and sea also led to more coasting vessels moving from Cape harbors to ports around the world. Exposure to this world was expressed in the probably fictitious story of a visitor in one of the earliest automobiles asking an old Cape Codder rocking on a chair at the general store if he knew the way to Sandwich. "No," the Cape Codder responded. "You don't get around much do you?" replied the visitor. "Oh, I've been to China, the West and East Indies, the Sandwich Islands but never made it to Sandwich town." Fishing, whaling, and merchant shipping, plus cranberry, turnip, potato, and asparagus growing as well as hay harvesting, sheep and cattle raising, and butter and cheese making, gave rise to comfortable homes and neat prosperous farms, whereas captaining, manufacturing salt, building boats, and merchandising gave rise to more substantial homes and estates.

These homes and estates, as well as shops loaded with goods from around the world and warehouses full of local products, signified the Cape's golden age. Young men and women looked forward to finding their livelihood either on the Cape or on the seas linked to the Cape and starting families. Provincetown Harbor in 1854 boasted one of the nation's largest fishing fleets, with over seven hundred boats including more than a hundred 90-ton schooners for fishing on the various fishing banks. Town harbors were busy centers of commercial activity. Typically there would be the saltworks, a bank, and an insurance company along with a warehouse, rope and sail works, and a shipyard for building and repairing ships and boats. Once a week a packet from Boston arrived, and dozens of boats based in the harbor left periodically to fish the banks.[99] Throughout the town there would be young people marrying, beginning families, and finding work. The little town of Wellfleet had ten school districts and several schools scattered about to serve the young of the community in the 1840s. Yet this golden age was not as pure gold as it seemed. Although more money was being made by Cape Codders, the environmental costs of making that money became increasingly evident.

RESOURCE FAILURE

Well before the exhaustion of the ocean's resources, the Cape's land suffered. The most obvious signs of resource failure were the disappearance of the region's forests, overgrazing on the dunes, and blowing sand. The problems of overgrazing on the dunes and decline of woodlands were not new to the nineteenth century: Cape Codders had been attempting, with town ordinances, to restrict grazing and protect their woods since before the American Revolution. In those days they had turned to petitions to the General Court. With both the economy and the Cape's population expanding, however, the problems became more acute. Woodlots shrank, and livestock numbers grew. Thus ensued an increased number of petitions to the legislature for help in protecting woodlands and dune grasses, "the principle defense and security of said meadows."[100] Despite the concern, the nineteenth-century state legislature was reluctant to step in to restrict private land use.[101]

When Timothy Dwight traveled across Cape Cod at the opening of the nineteenth century he noticed woods—not as thick as the Pilgrims had reported, but still woods.[102] A few years after Dwight, the Englishman Edward Augustus Kendall traversed the Cape and reported on the landscape "covered in wood."[103] A little more than fifty years after Dwight's and Kendall's travels, Henry David Thoreau made four trips to the peninsula in 1849 with the poet William Ellery Channing, and again in 1850 and 1855 and 1857, walking or riding a stagecoach across the Cape. Where Dwight had seen woods and fields Thoreau saw a desert. "The interior," Thoreau found, "was an exceedingly desolate landscape, with rarely a cultivated or cultivatable field in sight."[104] What most struck Thoreau was the absence of trees. In Truro "there was not a tree as far as we could see."[105] There were a few "thin belts of wood in Wellfleet and Truro, but for the most part we could see the horizon through them."[106] The Cape that Thoreau saw at midcentury, "nearly destitute of wood," was, he reminded his readers, only recently deforested. "All accounts agree in affirming that . . . the Cape was comparatively well wooded a century ago."[107] Where Thoreau found only tree stumps, he noted that thirty years earlier there had been "a flourishing forest."[108] "Large trees," Cape Codders reported to him, "once grew here." Instead of "the forests in midst of which they originally stood, barren heath with poverty-grass for heather now stretch away on every side."[109] And it was not only the Outer Cape that Thoreau found "vast and desolate." Dennis and Yarmouth, richly wooded in the eighteenth century, were "for the most part bare," with "barren hills" and "hardly a tree in sight."[110] By 1850 forests or woodlots covered less than 20 percent of the Outer Cape, a far lower proportion than for New England as a whole and a dramatic decline from the more than 80 percent that greeted the first white settlers.[111] Yarmouth's deforestation occurred relatively quickly in the mid-nineteenth century. The town still had significant enough woods as of 1843 to allow a fire sweeping through the town on May 12 to destroy four thousand acres of woodlands worth fifty thousand dollars. Even as late as 1854 John Sears built a steam-powered planing mill for finishing local lumber, but he shut it down in 1865 due to the lack of local lumber.[112] Truro residents told Frederick Freeman in 1860 that they remember their town being "clothed in wood."[113] When Edward Augustus Kendall traveled through this town in 1807, it "support[ed] lofty woods and hallowed into verdant and well watered vales," but it was, by Freeman's time, denuded of trees.[114]

The absence of trees and the overgrazing of the dunes also led to significant erosion with blowing sand covering once productive fields. Thoreau noted a field of 1,700 acres in Eastham that once grew wheat, but by the time of his visit "not now a particle of vegetable would grow."[115] Nor was blowing sand the only problem plaguing Cape farmers. Agnes Edwards, remembering the Cape of her youth, noted in 1918 that Cape farmers blamed the "sand" but they had themselves played a role in the soil's sterility. When farmers forced the soil, she claimed, "to bring forth crop after crop, all good was extracted and none returned and in the course of time it became utterly exhausted."[116] The combination of continuous plowing on increasingly tenuous land without sabbatical fallow, the destruction of the woodlands, and overgrazing on the fragile dunes turned the "deep and excellent soil" the Pilgrims had found into a "perfect desert of yellow sand" as Thoreau saw it, without "enough black earth to fill a flower pot."[117] Frederick Freeman noted in 1862 that the land "was gradually impoverished."[118] Eastham, once the breadbasket of the Plymouth Colony, continued to produce exportable amounts of grain into midcentury, but doing so became more difficult and the land paid the price. By 1860 over seventeen thousand acres of formerly fertile land was now useless. "Sand eroded from overgrazed and over lumbered tracts had blown over fertile field and made them barren."[119]

Sheep farming further exhausted the land. The rising price of wool encouraged the raising of more sheep and pasturing them on land too sandy to cultivate. There were large numbers of sheep by the early nineteenth century, especially on the fragile lands of the Outer Cape. Unfortunately for Cape shepherds, the solution to allowing sheep to run free on the dunes was to fence them in. Sheep fencing required four rails of cedar, and by the 1830s most of the Cape's cedar, which had once been common in the shallow kettle hole swamps, had been cut down and hauled away.

Other animals attacked the grasses that held down the sand. Many Cape families, whether or not they farmed, kept a milk cow, chickens, and a couple of pigs. Some families had a horse. In the winter, fodder for these animals came from the salt hay harvested across the Cape salt marshes, but, in the summer, the animals were put to pasture. The cheapest pastureland obtainable was barren sandy land, which was not cultivable and could be bought in the first half of the nineteenth century for as little as twenty-five cents an acre.[120] Beach grasses grew in this sandy dune soil, and animals could pasture during the summer months on these grasses.

The consequence was a disaster for the land. Strong winds blew across the Cape throughout the year, but especially in the months from late October until June when winds from twenty to fifty miles an hour regularly blew in from the sea. These were the winds that powered the windmills that ground the Cape's grain and pumped water for the saltworks. These winds also picked up sand—especially sand not anchored by roots or protected by thick growth of trees, bushes, and grass. The wind currents rose as they came off the coast and confronted the rise of the land, whether from the east or west, picking up loose sand as they rose and traveled across the Cape. Over land, the wind currents slowed so that as they reached the opposite shoreline, the sand fell. Land and shore on the leeward side gained sand. If that shore included a harbor, then as more sand was picked up by the winds, more sand filled in the harbor.

Blowing sand destroyed fields and filled in harbors. In the second half of the nineteenth century, drifting sand, as well as the natural migration of sand down from Provincetown, filled in Wellfleet's Duck Harbor, once a thriving center of eighteenth-century marine activity. Great Island, a prosperous whaling center for much of the eighteenth century, had been completely denuded of trees by the century's end and became a pasture for sheep, horses, and cattle. The combination of deforestation and overgrazing stripped the island of most of its vegetation and led to migration of its sand into Wellfleet Harbor.[121] Sand similarly threatened Pamet Harbor in Truro, while the town's northern East Harbor could hardly be navigated by the beginning of the nineteenth century. Chatham could no longer send "bankers"—fishing schooners that fished the banks—out to the fishing grounds because its harbor had silted-in too much for such large boats. For Chatham fishing the banks had been "greatly remunerative and did much to enrich the town." But by the mid-nineteenth century Chatham's fishermen turned to "coasting business," which involved vessels of "smaller draught," or mackerel fishing from smaller boats.[122] The most famous of the Cape's harbors—Provincetown's—was threatened with blowing sand as early as the late eighteenth century.

Dwight noticed, in his turn-of-the-century travels, that, although citizens were set to planting beach grass on the dunes around the harbor, the town still had more than 140 cows that wandered in search of food, "mak[ing] depredations on the beach grass."[123] In 1824 the state legislature appointed a commission to look into the degradation of the harbor. The commission reported that the 1786 act to protect the harbor was being neglected and "trees had been cut down for fuel and other purposes . . . the strip and waste depriving the sands of their natural protector." The sands, now "at the mercy of the winds . . . had been and continue to be forced over the meadows to their great detriment and also into the northeast part of the harbor."[124] The committee also reported that cutting trees and grazing animals on the dunes contributed to the erosion. It recommended a total prohibition of animal grazing on the dunes, along with outlawing the cutting of trees or beach grass, and it recommended planting of beach grass.

A state commission established in 1857 to investigate conditions of Provincetown Harbor once again noted that since 1714 the preservation of the harbor had been a major objective of the colonial and state governments.[125] Early attempts to protect the harbor from filling with sand involved limiting grazing on the lands behind the harbor, restricting the cutting of lumber and brush, and requiring residents to plant beach grass as their tax to the town. Despite these attempts, sand continued to flow into the harbor, and the lands behind it continued to deteriorate.

Blowing sand, declining production on agricultural land, and the destruction of the region's trees, along with the disappearance of the large schools of whales from the waters of the North Atlantic, did not bode well for the Cape's golden age. Truro, which in the eighteenth century was producing over twenty bushels of wheat and fifty bushels of corn per acre, was, by the nineteenth century, barely able to produce enough to meet its own needs.[126] By that time clams used for bait were producing a higher profit than cereal crops. Already by the start of the nineteenth century, clams dug for bait in Orleans yielded six hundred to one thousand barrels of bait clams per season. The poor particularly depended for their subsistence on digging clams to sell to fishermen.[127] By midcentury it was estimated that a thousand barrels of clams used for bait was equal in value to six to eight thousand bushels of grain, but even as Cape Codders were celebrating this bounty, shortages grew more acute. The small clam *Mya arenaria* was "not so plenty here as formerly."[128] The feather business, which produced fortunes in the early nineteenth century, ended in the 1840s, when the ducks had been killed off.[129]

A large number of Cape Codders were involved in boat- and shipbuilding in the early nineteenth century. Building boats was a highly skilled and labor-intensive occupation. A typical schooner was built from oak and pine and measured over 120 feet from stem to stern.[130] It would carry two or three masts with four to seven sets of gaff-rigged sails. A schooner going out to the banks in the nineteenth century would also carry ten to fourteen dories, which were also produced in local boatyards. Building these vessels consumed a great many labor hours as well as massive amounts of high-quality timber. As long as local supplies of good-quality oak and pine remained on the Cape, the industry thrived in almost every Cape Cod town. Shipwrights hired and trained local Cape Codders who built boats on the meadows and floated them out on the high tides. In 1837 the Cape produced $316,790 worth of ships and small craft. In 1845 Barnstable produced fifteen vessels, Falmouth eight, and Orleans six. Provincetown produced 150 ships, though they were mostly small vessels. Sandwich built a fifteen-thousand-dollar brig (a large square-rigged vessel) of four hundred tons.[131] But by midcentury local wood supplies began to dry up. Builders turned to expensive imported lumber from Maine and New Hampshire. But the cost of wood took its toll on the boatyards. By the 1850s the largest Cape vessels were no longer being rigged locally but towed to Boston for rigging. More and more yards closed. When the rigging left, so too did the rope making. Rope works that had been a major part of the early-nineteenth-century landscape and provided employment for many Cape Codders were abandoned. It became rare to see one.[132] In Harwich boat builders built their last schooner, the 70-ton *Job Chase,* in 1848. Truro, which had built some fifteen brigs and schooners (including the whaling vessels *Lydia* and *Sophia*) with local timbers over the first half of the nineteenth century, joined together her last large boat, the *Modena,* in 1850, also of local wood.[133] In Falmouth where shipwrights constructed large ships until the midcentury, boat builders shifted to smaller catboats, developed initially by the Crosby boat works. Thoreau reported being told: "Large schooners were once built of timber which once grew in Wellfleet. The old houses also were built of the timber of the Cape." By the time of Thoreau's visit "modern houses [were] built of what is called dimension timber imported from Maine, all ready to be set up."[134] Frederick Freeman noted in 1858 that "much ship-building was [formerly] carried on in different parts of the county, the supply of native timber being ample," but that, with the chopping of the region's trees, wood had to be imported from Maine. Dennis, the center of large vessel building on the Cape, turned out clipper ships into the 1860s, but produced fewer and fewer as the century moved into its second half.[135]

Although several factors, most prominently the decline of the federal tariff on salt and the influx of cheap solar-produced salt from Portugal, played into the dramatic decline of the saltworks. Notably, the dramatic rise in the cost of wood for construction of the salt vats contributed to squeezing the profits from the industry during the second half of the nineteenth century. The heavy storms that swept across the region, particularly in the winter, often broke up these vats that had been constructed along the shore. The gale of 1841 destroyed hundreds of vats.[136] The rebuilding of the vats required lumber that now had to be imported from Maine. Importing expensive dressed lumber dramatically raised the cost of salt production.[137] Saltworks also used cordwood for additional heat to reduce the moisture in the creation of Glauber's salts. The increased cost of this cordwood further hurt salt producers. With the lowering of tariffs on imported Portuguese salt, Cape Cod salt producers found they could not compete, and each decade from 1850 on saw fewer and fewer works. Provincetown, whose dozens of saltworks had proved profitable investments in the early nineteenth century, found the "high price of lumber," combined with the tariff reduction, too high a hurdle to continue.[138] By 1854 there were no more saltworks in the town.

Cape Codders had been cutting, building, burning, and clearing out wood since before the first Pilgrims moved down from Plymouth. Indian clearing of forestland in Barnstable was one factor that encouraged the Pilgrims to settle where they did. Throughout the seventeenth and eighteenth centuries, Cape Codders, both indigenous (in the early years) and European, puzzled over how to use the lumber resources of the Cape while still maintaining enough forest cover to protect the land and provide a continual resource for future use. It was not an easy process. Some felt the protections of the region's trees disadvantaged them by cutting off sources of income, and they resisted restraints on lumbering or using local woodlands. By the early decades of the nineteenth century, the dramatic increase in economic activity put more and more pressure on the woodland resources of the Cape; the age of ready supply of cheap or free lumber had ended. This in turn put additional economic pressure on two key Cape industries, boatbuilding and salt making. Excessive lumbering, increased grazing on delicate lands, and conversion of woodland and brushland into plowed fields and pastures also took their toll on the land and its ability to support sustainable agriculture.

5
THE DECLINE OF THE ESTABLISHED ECONOMY

The prosperity of the Cape's productive economy had not come easily. Although the fish abounded for decades, ships foundered in storms, accidents and injuries were commonplace, and disorderly crews could bring disaster. Clam digging and oyster farming demanded arduous labor, and Cape farming on sandy soils left crops prey to the uncertainties of weather. Yet hard work at sea and on land continued to bring Cape families a steady income until the natural resources they drew upon began to decline and falter. From 1860 to 1890 hard times appeared everywhere. The older economy of fishing, fish drying, shipbuilding, and salt making suffered serious decline. The cost of lumber had driven most large shipbuilders to relocate closer to sources of wood. The saltworks collapsed in the 1850s, not only because of the cost of wood, but also because of lowering of federal tariffs and declining demand for local salt for fish drying. Fishing boats increasingly avoided local harbors and took their iced fish directly to markets in Boston and New York. Local wharves were collapsing in disrepair. Only slowly did alternatives present themselves.

UPS, DOWNS, AND DIMINISHING RETURNS

While most of rural New England lost population during the first half of the nineteenth century, the Cape's population—like that of the nation as a whole

—grew dramatically in those years. At the beginning of the century, the population of Cape Cod was just over nineteen thousand residents. It grew to almost thirty-four thousand by midcentury. In the post–Civil War period, however, even as the overall U.S. population exploded with immigration, the Cape's population actually fell as more and more youths left for better opportunities on the mainland. By 1880 the Cape had only twenty-seven thousand residents; some towns lost as much as 40 percent or more of their population.[1] Wellfleet lost 57 percent of its population after the Civil War. From 1865 to 1895 the Cape as a whole lost over 20 percent of its population.

At the end of the eighteenth century—during the Revolutionary War and in the few years following—when the British blockaded the Cape and excluded Cape fishermen from the fishing banks off Canada, the region dropped into a severe depression. This led Fisher Ames to comment in 1789 that Cape Codders were "too poor to live there and . . . too poor to remove." Furnished with a healthy environment and economic base, the Cape quickly recovered from that depression. It was not so easy to recover from the decline that began in the middle of the nineteenth century. This decline came with the convergence of a number of factors both environmental and economic. They included shortages of wood, declining agricultural yields, smaller and shifting schools of fish, increased competition, and shifting technology and resource utilization. Unlike the Cape in the post–Revolutionary War period, Cape Codders were "too poor to live there" but not "too poor to remove."[2]

At midcentury Cape Codders who lived during the golden age that stretched from the end of the eighteenth century to the middle of the nineteenth were themselves hardly aware of the transition that would soon overtake them. The signs could be interpreted to suggest no major change was coming. There had been ups and downs in the economy even during the golden age. During the cold years of 1816 and 1817, when the Cape's climate was influenced by a volcanic eruption in Indonesia, Cape agriculture suffered. But the Cape fared better than the rest of New England, due to the moderating influence of the ocean, so Cape Cod farmers fared better than most during these two years.[3] Although the price for cereal crops failed to keep up with the ever more desperate struggle to get grain from sandy soil, asparagus, potatoes, and turnips did well, and the expanding cranberry market redirected farmers' eyes from failed cropland to previously wasted bogs.[4] The saltworks ceased functioning, but fishermen's growing need for ice led to increased harvesting of pond ice, which provided winter employment for those cutting, hauling, and storing this seasonal product.

Deep into the second half of the nineteenth century, the "haying grounds" of Barnstable's Great Salt Marsh were filled with "innumerable ricks" holding some eight thousand tons of hay used to fodder a growing number of animals, especially milk cows.[5] Cape Cod farmers believed that the future would be bright if they could effectively market what they grew. To that end they organized the Barnstable County Agricultural Society in 1844 and ran a county fair to show off their produce.[6]

In 1860 Barnstable County reached its peak of agricultural expansion, with 30 percent of the county actively farmed. Much of that expansion was at the expense of the forests that held the land against erosion.[7] But the farmers felt the lure of more cash from expanded production as well as the squeeze of competition from the far more fertile farms to the west. Even in the face of the negative consequences—especially, leaching out the humus from the soil—Cape farmers pushed their land past its limits.

The Civil War was probably even more a shock for Cape Codders than it was for much of the rest of the country. Being at sea or focused on the sea for long stretches of time, most Cape residents paid little attention to the growing political conflict about the nation. Some Cape Codders, especially the Quakers, took part in the campaign against slavery, but there were others who benefited from the prosperous southern trade and did not care to look deeply into the South's peculiar institution. Cape fish were traded to the slave plantations of the Deep South and West Indies, and cotton, spices, sugar, and molasses were brought back. Still, although ambivalent about slavery, Cape Codders did rally to the republic in wartime, not only swelling the ranks of the Massachusetts regiments but also, in particular, lending their seagoing experience to the Navy. As for most of the country, the war's end brought relief to many on the Cape; to many others it did not end the emptiness of loss, but Cape Codders were used to loss.

Peace saw them once again putting up sail for the fishing banks of the North Atlantic, scooping up cranberries, picking fruit, milking cows, and gathering up salt hay.[8] They did not return to seeking and slaughtering the leviathans of the deep.[9]

FEWER WHALES AND FEWER VESSELS

Whales were seldom seen now in the boreal waters north of Cape Cod. Although the number of whaling ships sailing out of Cape harbors declined each year as the century wore on, Cape Codders adjusted by adding more equipment and materials to their larger vessels and joining their Nantucket or New Bedford colleagues on the equatorial or Pacific arctic whaling grounds, or by switching from whaling to fishing.

The discovery of petroleum in western Pennsylvania in 1859 offered the nation an alternative for whale oil. Yet whale oil was still preferred over petroleum for lubricating machinery at a time when machine production was exploding around the country, and many felt it was superior to kerosene for lighting as well. Cape Codders abandoned the whaling trade well before the introduction of petroleum-based kerosene because the search for the ever-diminishing number of whales was far too costly in terms of time and money, and the catch far too meager to make it worth the long and dangerous journey. Whalers out of the Cape left the search for the ever-more elusive whales, and those remaining in nearby waters, to vessels that were better financed and capitalized sailing out of New Bedford and Nantucket.[10]

Like many Cape towns, Falmouth had a number of whaling ships sailing out in the early decades of the nineteenth century, but as the century progressed the amount of time these vessels remained at sea increased and the number of barrels of oil they brought home decreased. The Falmouth-built whaler *Awashonks* sailed out of Falmouth in 1830 and returned three years later with 2,000 barrels of sperm oil. When it left Falmouth in 1844 for the whaling grounds it stayed out four years and returned with 1,400 barrels of sperm oil and eleven barrels of other, less valuable, whale oil. The Falmouth whaler *Hobomok* left the harbor in 1836 and returned three years later with 2,000 of sperm oil and 1,000 barrels of less valuable oil. When it went whaling in 1856, it had to remain at sea for four years, returning with only thirty barrels of sperm oil and 1,572 barrels of other whale oil; this time it also brought home 10,500 pounds of bones as well. The 350-ton *Commodore Morris,* which sailed out in 1841, returned three years later with 1,400 barrels of sperm oil and forty barrels of other whale oil, but left the bones of the whales at sea. This ship went out again in 1858, it stayed five years and returned with 931 barrels of sperm oil, 232 barrels of other whale oil—and 1,700 pounds of bone which would have been left at sea fifteen years earlier.[11] In eighteenth-century Wellfleet, Barnstable, and Falmouth each had more than thirty-six whaling ships leaving their harbors for the North Atlantic and Chatham had twenty; the last whaling boat sailed out of Barnstable in 1846, and Wellfleet had but one left until it gave up whaling in 1867.[12] Three whalers struggled on in Provincetown to the end of the century.[13]

DIFFERENT KETTLES OF FISH

It was not just farmland and the whaling grounds that felt the pressure of heavy use. Fish stock, especially inshore fish species, experienced little-understood ups and downs. Sudden absences of fish from the shores, such as the dramatic decline of mackerel in the 1840s, engendered much anxiety and consternation. Yet, since the days when the first fishing line was lowered in the water, line and net fishermen have lived with the belief that tomorrow will bring new schools of fish. In the meantime they cast their lines and nets into schools of different fish—switching from cod to mackerel, for example—or they turned their attention to shingling a house, helping a neighbor build a barn, or picking up a job building a saltworks. Yet, even if the fishing for mackerel in Wellfleet dramatically declined, Provincetown was still booming. Cod, which in the eighteenth century had been available locally but was now mostly gone from local waters, could still be caught on the banks farther afield.[14] A visitor did note the presence, throughout the town, of flakes filled with drying cod.[15] Cod continued to be dried on flakes, and the introduction of ice in the 1840s for preserving the catch opened up new markets for fresh fish, particularly halibut and haddock.[16] The cutting and storing of ice frozen on the Cape's numerous ponds helped many a Cape family through lean winters. By mid-century herring fishing and drying had all but ceased in Provincetown, and its five smokehouses no longer gave off visible smoke. But river herring were still caught as they ran up the Cape's various streams and rivers. Spring herring runs provided a substantial amount of income for those involved in harvesting of this fish, as well as income for Cape towns leasing their herring runs. Between 1888 and 1890 fishermen on the Herring River in Wellfleet pulled in over two hundred thousand herring.[17]

Cod

Even though salt making, shipbuilding, and even fertilizer production were sliding into memory, the Cape still sent its youth to sea to catch cod. Despite its ups and downs, cod fishing could occasionally produce significant rewards, In 1885 William McKay brought in twenty-two thousand dollars in cod fish in one season, but such an event was rare after midcentury.[18] The cod fisheries along with other coastal commercial activity declined in the second part of the century.[19]

Like other Cape towns, Yarmouth, which had supported a healthy maritime center of fishing and coasting during the first half of the nineteenth century, saw its fortunes decline in the second half of the century. In the early 1800s Bass River, with its six wharves, had one brig that sailed to the West Indies and ten coasters that sailed from the Cape north to Boston, then south making stops in Connecticut, New York, and the southern states, and on to the West Indies. It had ten vessels fishing the Strait of Belle Isle, the shoals of Nova Scotia, and the Nantucket Shoals, and these were just the vessels leaving Bass River. By 1857 the *Yarmouth Register* noted that fishing had so declined that it "well nigh died out. Not more than two or three vessels have been sent from this port." In 1862, as the paper announced, "The last of the fishing fleet has been sold."[20]

Mackerel

Cape fishermen fished a variety of waters for a variety of catch. When the cod fishing was down, or there were no available berths in a cod boat, a fisherman would turn his attention to the shore fisheries. Mackerel were caught offshore by fishermen working on schooners, but unlike cod, large schools of mackerel also came close to the shore in the spring and again in the fall, allowing fishermen with hook and line working in dories and catboats to get to the fish.

Mackerel's importance to Cape Cod fishers grew over the nineteenth century. As early as 1826, Wellfleet shifted its focus from cod to offshore schools of mackerel. By 1851 the town had seventy-nine schooners with 852 men on board chasing mackerel. By midcentury Wellfleet was third behind Boston and Gloucester in its catch of fish.[21]

In the 1850s Chatham fishermen attempted to use an old Native American fishing method to catch schools of mackerel and herring as they swam along the coast. They pounded stakes into the soft sand and mud off the coast and stretched a net between the stakes to create a trap. Schools of fish would swim along the coast and come up against a line of netting running perpendicular to the shore. This would direct the schooling fish into a large heart-shaped netted pen. The fishermen then sailed or rowed out to the pen to scoop up the fish. These pound nets, or weirs, caught a variety of fish ranging from herring and menhaden to mackerel, bass, flounder, and bluefish.[22] Although local fishermen focused their attention on mackerel, during the 1880s bluefish was the more common catch. Occasionally, a large tuna would get caught in the net and wreak havoc. Until the advent of the twentieth-century tuna market, tuna were not a fish of preference but a nasty nuisance. The success of the Chatham experiment led other towns along the shallow Nantucket Sound and along the equally shallow north–south shore of Cape Cod Bay to set out pound nets in spring tides when the outgoing tide is at its most extreme. And the initial catch was impressive. A trap set up in Truro produced almost four hundred barrels of mackerel in its first season of operation.[23]

Pound nets proved profitable because they caught huge schools of fish. Although several factors account for fluctuating fish populations in any particular fishing grounds, hook and line fishermen, with some justification, blamed these pound nets for the declining schools of mackerel and menhaden.[24]

In 1853 Captain Isaiah Baker experimented off Chatham with a pursed seine net to catch mackerel. When feeding, mackerel travel in large schools close to the surface. The purse seine is a large light net, the lower end of which can be pulled together like a purse. After sighting a school of mackerel, a dory was lowered with one end of the net on board the ship. The fishermen rowed the dory around the school of mackerel, letting out the net as they went. Once the fish were surrounded, the net was pulled in and the fish caught. Purse seines, when successful, brought in huge hauls, but the process was difficult to carry out. Mackerel swim fast, and, if the seine fishermen were not both lucky and careful, a potential harvest of over one hundred barrels of fish would quickly dart away, leaving the fishermen with an empty seine. Because the new technique was not only riskier, but also more capital-intensive than hand lining, many Cape fishermen were unwilling or unable to switch to purse seine fishing. As late as 1889, when the catch from seine fishing vastly outstripped that of hand liners, Wellfleet still sent out eight hand-line fishing boats along with its thirteen "seiners."[25]

The habit of mackerel, to school near the shore when searching for food, allowed fishermen with smaller craft to still bring in a catch. The silting-in of Chatham's harbor forced its fishermen to switch to smaller boats that could negotiate its shallow waters. Although these boats were not capable of fighting the rough seas of Georges Bank, they could search out mackerel close at hand.[26]

Dependence on mackerel fisheries to make up for the slack in the cod catch proved to be an unfortunate bet. The railroad may have made it easier for Cape oystermen and cranberry farmers to get their goods to market, but the railroad also integrated the larger national economy. Among many other effects, Great Lakes' white fish could now be offered to East Coast fish eaters, dampening the demand for the more oily mackerel. The adoption of Isaiah Baker's purse seine also reduced the numbers of youth who could find a berth on a hook and line operation.[27] Although its importance had risen by midcentury, mackerel fishing declined in the 1880s.[28] The mackerel catch of 1890 was 10 percent down from what it had been a few decades earlier.[29] Orleans, for example, formerly had three companies buying and processing mackerel from a large number of local fishermen, but by 1885, it had no mackerel fishing.

Further north, during the first half of the century, young men from Wellfleet and Truro had fished for cod on the banks and for mackerel offshore. With the decline of offshore mackerel and greater difficulty in catching a profitable load of cod, fishermen from both towns found weir and trap fishing "more profitable." But more profitable didn't mean prosperous. Truro's population in 1850 was 2,051, but it fell to 1,269 in 1870 and to 972 in 1885. Across the Cape, investments in fishing declined rapidly, as did fishing income. In 1885 Cape Codders put almost two million dollars in fishing but by 1895 that investment was down to just under seven hundred thousand dollars. The value of fish caught had declined 40 percent.[30] And it was not only fishermen who were hurt with the shrinking catch. As investigators for the Bureau of Statistics noted in 1897, "The industry also gave employment to many men packing and handling the large quantities of fish brought in to their wharves and to many mechanics who were employed on the vessels in making repairs, such as ship carpenters, painters, caulkers, riggers, block makers, and sail makers."[31] With less fish these people were also without work.

The hundred ships that left Wellfleet Harbor in 1850 declined to only twenty by 1890, and these were "chiefly of the smaller class." The hundred fishing ships of 1850 (along with the town's shellfish harvesting and what was left of its whaling fleet) had supported a population of 2,411, including Lorenzo Dow Baker, but its fish catch fell 76 percent just in the last two decades of the nineteenth century.[32] The Wellfleet shipbuilder Edwin Rogers left the town in the late 1850s when over eighty sailing ships filled its harbors. When he returned in 1895 the harbor was almost empty: "Large fishing wharves [were] unoccupied and fast falling into decay; a fourth wharf had entirely disappeared. Two abandoned homes of fishermen were in sight, whose roofs had fallen in and whose windows and doors were shattered or unhung and these ruins told the story of many others."[33] By 1885 only 1,687 people called Wellfleet their residence, although many more who were scattered about the country still called it home.[34]

BOATBUILDING FROM SCHOONER TO SKIFF

The passing of the amphibious age was reflected in the dramatic reduction in boatbuilding. While the increasing costs of imported lumber squeezed Cape Cod ship- and boat-wrights, the large pool of skilled boat builders and the existing ship- and boat-works infrastructure kept the industry alive into the postwar period. Barnstable, for example, with eleven boatyards produced seventeen boats in 1885 compared to fifteen in 1845. But the boats they constructed were small, with a total value of only $6,377. Provincetown built 150 boats in 1845, but in 1885 it produced only 39 with a value of only $6,800. Building sailing vessels, which had been a "staple industry" of Cape towns in the early part of the century, became, in the postwar period, less and less a part of the Cape's coastal landscape.[35] Boatbuilding had always been dependent on lumber and skilled workers, but also on demand, and demand was dependent upon a variety of factors: ships lost to war, ships lost to storms, changing sailing technology, anticipated increase in trade, increased access to credit, higher prices for goods shipped by sea. Because of these factors, demand fluctuated even during the golden age, but when demand was high, builders and shipwrights tried to forget the lows and focused on completing the vessel in the yard.[36]

Unfortunately by 1870 it became harder and harder to keep the optimism. The October gale of 1841 destroyed many of the Cape's shipyards. With the increase in the price of wood for building boats many of the smaller operations could not restart. The larger ship works, with their pool of skilled workers, access to credit, and strong reputations, continued to survive. Edwin Rogers told an investigator for the Massachusetts Bureau of Labor Statistics that he was still building boats in 1850 for the mackerel fisheries and employing a number of local men in the process, but soon afterward he abandoned Wellfleet and boatbuilding.[37]

The sixty dollars per ton freight rates for shipping to Europe at midcentury and the discovery of gold in California kept the demand for ocean-going vessels, particularly the fast new clipper ships, high. The Shiverick works in Dennis not only continued to build schooners but also built these famous clipper ships. After the 1841 gale the yard used mostly lumber imported from Maine.[38] Between 1849 and 1863 it built eight ships of more than a thousand tons, as well as four large schooners.[39] But even the Shivericks felt the pressure of rising costs for wood. They launched their last clipper ship in 1863.

Although in the first half of the century there were active boat works in most Cape towns employing teams of workers, by 1885 only Barnstable and Provincetown reported boat builders to the state census takers, eleven in the former town and twenty ships' carpenters in the latter.[40] The combination of unstable demand and dramatically increasing expenses for materials that were no longer cheap or locally available overwhelmed the ability of skilled labor and infrastructure to hold more than a fraction of the industry on the Cape. Those boat works that did survive focused their attention on the construction of smaller sailing skiffs—particularly catboats, which Jesse Crosby developed in 1845 using local pine, oak, and cedar. These smaller craft were still in demand by coastal fishermen, oystermen, lobstermen, and scallopers, but also by a new species to the Cape, the summer visitor.

ABANDONED SALTWORKS AND DEMOLISHED VATS

Boatbuilding was not the only Cape industry to be squeezed by declining demand and rising costs of lumber supplies. Most of the two million dollars invested in saltworks across the Cape at midcentury represented the vast array of windmills, wooden vats, and vat roofs that spread across the shoreline from Sandwich and Falmouth to Provincetown. Typical was Wellfleet, which had thirty-nine solar saltworks in 1837 producing over 10,000 bushels a year.[41] Initially the wood for these structures, wide board of pine, came from local forests, but by midcentury it had to be imported from Maine. The system of windmills, vats, and roofs needed constant repair and rebuilding so that each year those costs rose while competition from other sources of salt cut into profits.[42] Barnstable had twenty-four saltworks in 1845 producing over 21,000 bushels of salt, but by 1855 there were only eleven producing 10,550 bushels of salt. In 1865 Barnstable had only three saltworks producing a little over 3,000 bushels. Brewster had thirty-nine plants in 1845, but only twelve remained twenty years later. Dennis had eighty-five saltworks producing almost 35,000 bushels of salt in 1845, but by 1865 it had twenty-three works producing 15,275 bushels, and in 1885 only one lonely manufacturer continued, producing only 300 bushels of salt. Provincetown had seventy saltworks in 1845 that produced 26,000 bushels, but in 1865 it had one saltworks producing only 200 bushels. Yarmouth, one of the largest centers for salt production, had sixty-five saltworks in 1845 producing over 74,000 bushels of salt. In 1865 only nineteen remained producing less than 14,000 bushels. Ten years later only three held out, and by 1885 only one remained, producing primarily Epson salts.[43]

By the 1870s and 1880s, rather than repair saltworks, owners sold off the wood from the old vats to other Cape Codders who used it to repair their homes or barns. Salt-vat wood was considered good barn material because the salt protected the wood from carpenter ants and termites. But for the Cape Codder who worked in the saltworks, dug clams to sell to fishermen for bait, and raised chickens and vegetables to maintain his family, the demolition of the old vats represented an end to a way of life. The family might be able to continue digging clams, selling eggs and vegetables to the local market, or picking up work on one of the coasting fishing boats, or the wife might find employment as a servant or cook in hotels or the homes of others, but without the steady income from the saltworks, it was increasingly clear the children would have to look elsewhere for employment and a future. And that elsewhere was not at the local boat works or even at the docks loading and unloading the packets.

Cape Codders no longer repaired saltworks or found work in boat works, but neither did they spend their days building houses, barns, or warehouse. As a Wellfleetian noted at the end of the century, "no new house had been built there since anyone could remember."[44]

PACKETS, THE RAILROAD, AND THE END OF THE AMPHIBIOUS AGE

Whalers and fishermen were not the only seamen who found it difficult to continue their traditional work. On May 28, 1848, Sandwich residents celebrated the coming of the railroad, the symbol of nineteenth-century progress. Now goods from the western Cape were no longer dependent upon sailing ships to get to market, and passengers were no longer concerned about the uncertainty of weather, which could transform a six-hour trip to a two-day trip. The modern miracle of steam and rail would overcome the vagaries of shifting or dying winds, and stories of long, rough crossings became the tales of old timers. Success of the Cape Cod Branch, which brought the railroad south from Middleboro to Sandwich, laid the groundwork for the extension of the line on to Hyannis and ultimately to Provincetown.[45] The laying of

The Decline of the Established Economy

the tracks, however, proved problematic: sand made a poor rail bed. It took another six years for the line to finally arrive at Barnstable in 1854, and six more years to connect to Hyannis and the steam ferry that by then was running from Hyannis to Nantucket and Martha's Vineyard. Each year the rails stretched farther out on the Cape, reaching Orleans in 1865, Wellfleet in 1870, Provincetown in 1873, and Chatham in 1887. As the service became more regular, fewer and fewer packet boats left Cape harbors.[46] Simeon Higgins, who captained a packet schooner out of Orleans from the 1820s to the 1870s, abandoned his schooner to become a hotelkeeper.[47] Higgins was not alone.

Trains were not the only form of steam locomotion pushing up against the Cape's traditional culture. Steamboats increasingly replaced sailing ships as a more dependable means of moving goods across the water, just as the steam ferry replaced the traditional sailing packet serving Provincetown Harbor. But steamships required deep harbors. With the exception of Provincetown most Cape Cod harbors were shallow. The combination of the increased dependence on steam ferries and the use of the railroad undermined the local packets that served the small ports along the Cape. The slow replacement of steam vessels for sailing vessels also led to the decline of economic activity around the harbors themselves. Stores closed or moved off the wharf into town to be closer to the train depot. Sailmaking lofts emptied of sail makers, coopers looked elsewhere to make their barrels. Rope works grew over with beach grass.[48]

The coming of the railroad to the Cape's tip effectively ended the packet service with the exception of service from Provincetown and Wellfleet to Boston, which remained, greatly diminished, until the end of the century. Even these runs were impacted by the age of steam. In 1885 the 500-ton steamer *Longfellow* offered service between Boston and Provincetown and soon came to dominate the link between the Outer Cape and Boston.[49]

Bulk items, such as lumber or coal, continued to be moved off and on the Cape by boat, but the dozens of Cape sailors who sailed the packets found fewer and fewer berths from which to work. Most of the Cape packets were built in Cape boatyards. Orders no longer came in for new schooners for the packet runs. The old packets were offered up as fishing boats. With the passage of the packets into historical memory, the Cape's amphibious age became more a part of the region's lore than of its present reality. By the 1870s, Provincetown, which did not have streets until 1838 because it was easier to move things by boat than by wagon, found that its train depot was as much a center of activity as its wharves.

IRON RAILS, OVERLAND COMMERCE, AND THE CHANGING ECOSYSTEM

Although the railroad slowly brought an end to the packet service, it also brought new industry or change of activity as well. In 1864 the Keith Car and Manufacturing Company, founded at Sagamore, began building and repairing freight cars for the railroad. The company, which manufactured Conestoga wagons before switching to the more profitable railroad cars, was, at one point, the largest employer in the state, hiring more people than the more renowned textile companies to the north. While the Cape Cod boat builders and salt makers saw their industries disappear in the second half of the nineteenth century, others saw only progress as the Cape's future. The coming of the railroad meant that goods to and from the mainland could move more quickly and regularly. And one of the major commodities moved off the Cape by rail was the cranberry.

From Marshland to Bog

Cranberries were a perfect Cape crop, for not only did they grow prolifically on heavily modified bog-land, but they also kept well without refrigeration over a long period of time. When processed with sugar or honey they made a tasty vitamin C–rich food, especially for sailors long at sea. The conversion of marshes to cranberry bogs went ahead, seemingly as fast as the new trains could take the crop to market, driving up the price of previously worthless land.

Although opening up a cranberry bog was difficult and expensive, the decline in the price of sugar after 1845 gave cranberries a significant boost in popularity, encouraging even larger operations. By midcentury Nathan Smith had hundreds of acres in cranberries and employed more than fifty pickers at harvest time.[50] In 1859 an acre of cranberry bog was worth ten thousand dollars, an astounding price that matched the price of land in Boston. In that year cranberries worth almost twenty-eight thousand dollars were sent to market, while thirty years later 150,000 barrels worth over a million dollars moved off-Cape.[51] Even small operators became involved with cranberry production. Katherine Lee Bates remembered getting through the hard times of the second half of the nineteenth century by working a small two-acre cranberry bog along with a home garden, some chickens, and a cow.[52] With the expansion of cranberry production came the need for barrels. Although barrel making was not as

The Decline of the Established Economy

common in Cape towns in the post–Civil War period as it had been, local demand by cranberry growers kept many of the larger factories running. The Cahoon Barrel Factory in Harwich switched from general barrel making to the exclusive production of cranberry barrels.[53]

Cranberries were not the only Cape products moving to market on iron rails. Even Wellfleet oysters moved away from the sea on their way to market. By the 1870s Wellfleet was shipping railroad cars of oysters to New York and Boston markets. So lucrative were these oysters that the railroad built spur lines right to the oyster shacks.

Estuarial Habitat among Dunes and Dikes

The railroad industry ultimately conquered the shifting sands by building dikes and elevating the beds for rails. The building of the railroad was expensive, but the railroad boosters convinced the people of Cape Cod that their futures would blow away like the sands upon the outer dunes without the railroads. Cape Codders responded. Much of the capital to build the Old Colony Line was raised locally with families buying a few shares in what they believed was the future of their communities.[54] What they did not fully understand was that it was a future very different from the past they knew and thought they were protecting.

The very building of the railroad itself transformed the natural landscape of the Cape far more than just the appearance of rail lines and the smoking engines. The Cape Cod Branch Rail Road's dikes were more massive and successful than the older dikes (built to drain marshes), in blocking tidal flow and transforming salt marshes to brackish marshes. In many towns the dikes cut off or greatly restricted avenues for migrating fish. The dikes also changed the ecosystems of the estuaries. The saltwater marshes that wove through most of the Cape towns not only supplied salt hay to farmers, but also were nutrient generators for complex estuary systems. They served as nurseries and breeding grounds for fish, and the predominant eelgrass provided food for multiple small creatures and the base for developing scallops. The tides that swept through the marshes brought nutrients into the bays and harbors of the Cape enriching the shellfish grounds. Saltwater marshes are among the world's most productive environments, but construction of the railroads dramatically reduced the flow of nutrients through the system and created stagnant waters that were no longer flushed out with the tides.[55] Eventually phragmites (reeds), cattails, and other freshwater plants, as well as fauna and flora new to the region came to reside behind the dikes.[56] In a report on the causes of the depression on the Cape, investigators for the Bureau of Statistics of Labor noted in 1897 that the effect of a dike built earlier in the century as a roadbed on the salt meadow was "unfavorable as this meadow is now overrun with flags."[57]

Glass and Guano

Although Cape Cod was an amphibious culture during the first half of the nineteenth century, with well over 60 percent of its income coming from fishing, shellfish harvesting (clams and oysters), whaling, coasting, or manufacturing goods that supported those activities, Cape Cod also supported a significant manufacturing enterprise that had nothing to do with the sea. It was also tied to local supplies. A glass manufacturing plant had been established in Sandwich in 1825 to take advantage of the ready supply of marsh grass for packing and shipping and the large forest of pine that stretched from Sandwich to Falmouth. This was no small works. Incorporated in 1825 with three hundred thousand dollars in initial capital, the Boston and Sandwich Glass Company was the nation's largest glassworks.[58] The company bought over two thousand acres of forest to use for fuel. At its height it had its own wharf and its own steam and sailing boats to bring in materials; eventually a rail line linked the wharf to the factory and another connected with the main rail line to Boston. By 1850 the firm employed some five hundred workers. Before the works shut down in 1894 it produced over thirty million dollars worth of glass. But by the late 1880s the supply of cheap fuel was running out and competitors distant from the Cape were producing colored "Sandwich Glass." Squeezed by rising costs for materials and fuel (wood from the company forest), the company cut wages. A bitter strike followed and the company closed its doors for good. Locals attempted to keep glass working in Sandwich, but their efforts only slowed the final end of glass manufacturing on the Cape.

Glass making and railroad car manufacturing were not the only manufacturing enterprises on the Cape. Hoping to take advantage of local fish stock, particularly the thick schools of menhaden that swam into Cape waters each summer, a fertilizer company was established on an abandoned wharf in Woods Hole. The company hoped to import guano from the Pacific South America and mix it with ground-up menhaden to sell to off-Cape farmers as fertilizer. The company prospered for several years until the declining menhaden catch

The Decline of the Established Economy 95

forced it to focus more heavily on guano. Eventually, and fortunately for those like Ezra Perry who hoped to convert the old whaling and shipbuilding station into a fancy summer resort, it closed its doors.

THREE VISIONS FOR A NEW CAPE COD

At the beginning of the twentieth century three Cape Codders who had been central participants in the regime of extraction and production realized that the Cape Cod of fishermen, sailors, and farmers was ending, and a new era was being born. One of those people, whom we have met before, was the famous banana captain Lorenzo Dow Baker from Wellfleet, the town situated on what the locals called the Lower Cape. Another was Ezra George Perry. As a young man from Bourne on the Upper Cape—the southwestern end or upwind part of the Cape—Perry left the seaman's life behind and migrated to the mainland to make a new life in the city. The third was Nathan Crosby.

Crosby: Returning in Retirement

Nathan Crosby began working as a tanner in Chatham, but left that town in 1835 to return to his family farm in Brewster, east of Dennis and just west of where the Cape bends northward. Here he invested in several fishing boats.

But like thousands of his fellow Cape Codders, Crosby came to believe there was no future on the Cape for tanning, fishing, or farming. When his fishing investment turned sour he left the Cape for economic opportunity elsewhere, striking out in 1854 for the rapidly expanding western city of Chicago, where he opened a brewery and prospered as a businessman. In 1882 he turned the business over to his son and retired to the family farm in Brewster, where he died unexpectedly. His son, Albert, grew rich in the brewing business. In 1887 Albert also retired and returned to Brewster, where, in 1888, he built a huge estate called Tawasentha.[59]

Although both father and son saw their move back to Brewster as a return home, it was a very different home they returned to. The Crosby legacy also contributed to that difference. Within a few years of being built, the Crosby estate, Tawasentha, was made over into a tourist hotel.

Perry: Outside Money and Local Color

Ezra George Perry, the Upper Cape migrant to the city, had a hard youth. His father, a sea captain, was killed at sea by a mutinous crew. His mother managed to hold the family together by selling off bits of the family farm to other town farmers. Perry's brother went to sea to help support the family. When Perry reached his sixteenth year he also packed his seabag and took his place before the mast. Unlike his brother, Perry did not see a future as a sailor. After four years he traded in his sea legs, first for a job with the new street railroads in Boston, then as a jobber selling confections. Perry did well as a salesman, and was soon selling real estate, but he missed his home on the Cape. His experiences in Boston convinced him that money could still be made on the Cape, not from fishing and farming but from selling the Cape itself to the wealthy of the city. At the end of the nineteenth century, Perry came back to Bourne, his hometown, where he began buying up pastureland around Phinney Harbor and developing it into lots to be sold to the urban rich for seaside summer estates.[60] Perry knew his audience and knew what they wanted. He produced glossy travel books about Buzzards Bay, his home area, and Cape Cod itself. These books were guides to Cape towns, and they focused on the quaintness of the area, its healthy climate, its churches, and its libraries. They also focused on how other persons of wealth, such as President Grover Cleveland, were buying land along the coast of the Cape and building substantial summer estates and institutions like golf courses and yacht clubs to meet their cultural, recreational, and social needs.[61]

Waterfront property was relatively cheap when Perry and others began buying up land.[62] Perry envisioned a new Cape Cod whose economy would be driven by outside money and where Cape residents would provide local color, labor to maintain and serve on the estates, and local produce, dairy, and meat products for summer residents hungry from busy days of tennis, yachting, sport fishing, and golf.[63]

Baker: The Value of Community

Lorenzo Dow Baker, like Perry, foresaw a Cape economy resting on tourists and summer visitors who would appreciate its quaint, picturesque villages and the stunning beauty of its fragile landscape. Baker's vision was shaped not only by his belief that the older Cape economy was dying but also by his experience growing up on the Cape. Baker was a Methodist. In his youth he attended the Methodist camp meetings that drew thousands of enthusiasts each summer, first to Eastham and then to Yarmouth. Initially, these worshipers bedded down in tents, but during the late nineteenth century, the congregants built small cottages on the campgrounds and little villages emerged. As the camp meetings became less focused on evangelicalism and more on community, fellowship, and enjoyment of time by the sea, it became the custom for church members to come each year and bring their entire families. By the end of the nineteenth century campers were extending their stay beyond the revivals for family time away from the crowds and pollution of the city.[64] Baker was a regular participant in the camp meetings and had his own cottage at the village in Yarmouth. He understood the appeal of a peaceful seaside retreat for ordinary folks like those who gathered there each August.

Baker also understood how to organize large systems. The banana trade needed a massive infrastructure to sustain it and Baker helped create it. Baker brought his camp meeting experience and his work in transforming Port Antonio, Jamaica, to his hometown of Wellfleet. He built the Chequessett Inn on the old wharf in Wellfleet; it was a success, but it was only part of Baker's vision for a different Cape Cod. Baker also bought up land and built small summer cottages to rent out to middle-class urban Americans looking to escape the dirt and grime of the city.

VACATIONS FOR THE FEW OR THE MANY?

Although Perry and Baker shared the belief that the future of the Cape lay in the summer visitor, they had different visions of how that would play out. And the different villages of the Cape reflected that difference. For Perry the summer visitors were the family and friends of the very wealthy. His new Cape Cod was a Cape Cod of summer estates, grand hotels, lavish golf courses, yacht clubs, lawn parties, and croquet. But Baker saw the new Cape as an extension of the Methodist camp meeting, a place of community, fellowship, and peaceful contemplation as well as a place for vigorous wading, rowing skiffs (rather than sailboats and yachts), and fishing. It would be a place where people would spend money and others would make money, but it would be the money of the many, and the many were the urban middle class. Both Perry's and Baker's visions of the Cape came into being in the early twentieth century. The shoreline of Buzzards Bay and the south shore from Falmouth to Hyannis began to fill up with large estates and grand hotels, as did eastern Harwich and Chatham. The shore from Yarmouth to Harwich and the bay side from Eastham to Truro saw a proliferation of small cottages and summer cottage villages.

Perry's and Baker's visions of the Cape overwhelmed traditionally conservative Cape Codders because the two-hundred-year regime of resource extraction had driven the Cape, especially since about 1860, into an environmental crisis that brought on an economic collapse. The thick forests had been stripped from the land. Eroded soil and blowing sand plagued Cape farmers, who also struggled with low prices and constrained markets. The sea's fecundity, once seemingly endless, was increasingly in doubt.[65] The salt marshes, crucial as fodder for livestock for almost three hundred years, now seemed of little importance in a world of steam engines. Ezra George Perry and Lorenzo Dow Baker believed the key to the recovery of the Cape's economy and its environment lay in the new economy of tourism. Farms would become estates, while shoreline saltworks, fish flakes, and fish houses would become vacation cottages and bathing beaches. Pastures would become croquet fields and tennis courts. Fishing piers would become docks for pleasure boats and sport fishing vessels. Local farmers would produce food for vacationing visitors while their children would become homebuilders and gardeners.

FROM CRISIS TO TRANSFORMATION, STEP BY STEP

The transition from an economy of production to one of tourism, although envisioned in smooth and idyllic terms by both Perry and Baker, proved neither smooth nor idyllic, nor was it quick. By the middle of the nineteenth century it was common for the wealthy to escape the tensions and congestion of the city and appreciate what Henry David Thoreau called "all out of doors, huge and real."[66] For the most part, well-to-do folk who sought out recreation and relaxation in the middle decades of the nineteenth century did not look to Cape Cod for their non-urban comfort.[67] Thoreau felt that "the time must come when this coast will be a place of resort for those New Englanders who really wish to visit the seaside," but at midcentury it was still, in Thoreau's words, "wholly unknown to the fashionable world."[68] Although Thoreau depicted the rugged beauty of the Cape, his Cape Cod work was little read.

The World of Wealthy Sportsmen

The exceptions to this isolation were men drawn to the Cape for hunting and fishing. Before Lorenzo Dow Baker took to building his vacation cottages and Perry began buying up land to develop, sport hunters and fishermen who lived elsewhere came to the Cape, not to work but to recreate away from home. Daniel Webster and the natural scientist Theodore Lyman (who was also a U.S. representative to Congress in the mid-1800s), recounted to their friends in Boston that the Cape's rich supply of fish and fauna provided great sport for those willing to make the trip.[69] Once the railroad brought the Cape within a half-day's journey of Boston, hunters and fishers began taking the train down for a week or weekend of sport. Fishing and hunting gained in popularity not only for the great fun of the sport, but also for serving as a measure of manliness to the increasingly urban upper classes.[70]

The decline of agriculture left more fallow land and more habitats for game animals. Although the commercial fish stocks targeted by Cape Codders were in steep decline, the fish preferred by sportsmen still abounded. Ponds covered over one third of the Cape, and many had thriving populations of bass, perch, and other fish cherished by sportsmen. Soon hundreds of men were taking the train down and trekking off to their favorite fishing hole, while others climbed across dunes to try their hand at catching a striped bass or bluefish. The 1890 history of Barnstable County noted that although commercial fishermen saw bluefish as a nuisance fish, sports fishing had made the bluefish "among the favorite sea fish," and catching them became a "standard sport with pleasure seekers."[71] As soon as the railroad reached Sandwich, those interested in catching fish or shooting ducks, geese, and ground birds—and financially secure enough to be able to leave Boston or New York for a weekend or longer—began coming to the Cape for sport fishing and hunting.[72] They bought up land along the ponds or streams and had rough camping cabins erected. But these cabins were simple affairs and occupied for brief periods of time.[73] Sport hunters and fishers also established sports clubs, which bought land and erected clubhouses and rustic hunting shacks. These private camps were looked after by locals who would also arrange to pick up the sportsmen at the railroad depot, get them to the camp, and supply food.[74] For these midcentury outdoorsmen, the Cape was not a place to bring family or settle in for the season, but a temporary refuge from responsibility and urban pressures. But, with increasing frequency, men chose to bring their families along.

Like many other upper-class New England young men, Loring Underwood loved to hunt and fish. While an undergraduate at Harvard he regularly took the train from Boston to the Cape to pursue both interests. By the end of his time at Harvard, Underwood secured property in Chatham and constructed a simple hunting lodge. Separated by a difficult older brother from the family business of canned goods, most notably deviled ham, Underwood focused his interest on landscape architecture. Although photography and landscape architecture occupied Underwood's time, hunting and fishing were his passions. With the leisure that family wealth provided, Underwood spent more and more time in Chatham. He married Emily Walton shortly after his graduation from Harvard in 1897, and, after touring and studying under Édouard André at the École nationale supérieure d' horticulture in Paris, he returned to the United States and his frequent hunting trips to the Cape. In 1913, to accommodate his growing family, he built a more substantial summer cottage in Chatham where the family could enjoy the benefits of the wider range of summer activities.[75]

The Revival Camp Meetings

Besides wealthy hunters and fishers, the first group of pioneer visitors who came as families to the Cape were those who came to what Thoreau called "the best place of all our coast to go to," for the comforts of the soul and spirit that was offered by the Methodist revivals.[76]

The first camp meetings began in August of 1819 in Truro, but moved shortly to Eastham in 1838 with the incorporation of Camp-Meeting Grove Corporation. The corporation held a tract of land reserved for the annual religious meetings and brought thousands of faithful to Millennium Grove for the revivals.[77] Initially the revivalists came by boat from the surrounding Cape towns. By the 1850s, however, the meetings drew in revivalists from off Cape. The Methodists gathered in tents for these weeklong retreats. Soon the less rugged congregants built small, simple cottages at the edge of the tents. As the numbers and the distance from home grew longer, the logistics of moving thousands of people in and out of Eastham by packet boat proved difficult. When the railroad reached Yarmouth it was decided in 1863 to move the revivals to the new Millennium Grove a mile and a half from the Yarmouth train depot. The new location offered more space and was far more convenient, especially for those coming from far off the Cape. Soon a wooden tabernacle

The Decline of the Established Economy

At the end of the nineteenth century, upper-class male New Englanders, like those at this duck blind, established hunting and fishing camps and cabins on the Cape as retreats from the pressures of urban life. Increasingly these cabins and camps were enlarged and converted into family retreats. Photo courtesy of the Underwood family.

replaced the open-air meetings and cottages replaced tents. It became "the fashion for good church members to come every year bringing their entire families," one old timer remembered.[78] As the cottages became less rugged it became the fashion for church members to stay after the revival to appreciate the cool summer air. Even before the move to Yarmouth, Thoreau noted that although the preachers drew people to the camp meetings, the ocean was also an appeal: "They all stream over here" to the ocean during their stay."[79] Some who built a cottage at Millennium Grove rented out their cottages to other church-affiliated families before and after the revival so that they might also have the opportunity for a seaside escape.[80]

The success of Millennium Grove encouraged other denominations to open revival camps on the Cape. In 1872, Christian Churches of Southeastern Massachusetts established a camp in Craigsville, a village in southwest Barnstable, which soon had cabins and two hotels catering to those coming for spiritual and emotional healing by the sea. In addition to the camp meeting, the Christian Churches purchased abandoned saltworks on beachfront property that ultimately became Craigsville Beach. By the 1870s these camp meetings were drawing between three and seven thousand visitors a day throughout the month of August.[81]

Unlike the wealthy whom Perry was attempting to lure to his developments on the Cape, those who came to the Methodist revivals were of the middling sort. A few of the better off, such as Lorenzo Dow Baker (who continued to be a regular participant even after he became wealthy), were represented at the camp meetings; their cottages, although modest, were usually larger and more elaborate than those around them. Yet the typical participant was a small businessperson, an owner of a retail store, a clerk, or a skilled artisan. For these members of the growing middle, lower middle, and upper working classes, respectability and hard work were core elements of their belief system. They retained a healthy suspicion of relaxation and leisure, regarding them as the vanities of the very wealthy. Coming to the Cape for a revival was as far from going to frivolous Newport as they could imagine. Staying on to relax in the clean, wholesome air among fellow churchgoers was also imagined to be radically different from the behavior of the less temperate well-to-do visitors.

Although the revivals continued to bring the faithful to the Cape into the twentieth century, in time it was the experience of being on the Cape itself that drew more and more people to the simple cottages. By the twentieth century the cottages and patterns of retreat of the Methodist revivalists became the model for an ever-larger circle of the expanding middle classes.

Returning to the Family Home

Methodists were not the only pioneer vacationers to see the Cape as a place for long family visits. These other vacationers were not the fashionable city dwellers so much as Cape Codders who had moved off-Cape to make their fortune and returned home in the summer for visits to the old homestead; or who, like Nathan Crosby, had made their fortunes elsewhere and then returned to the Cape to retire. These diaspora Cape Codders were the vanguard of what would become a steady stream of visitors. The Cape had been losing its youth to the greater opportunities of the mainland since midcentury. Although few were as successful as Nathan Crosby or Gustavus Swift, the butcher from Eastham who founded a meatpacking empire in Chicago, many of the "young men and women of the village who had their living to earn were obliged to remove to large cities," for stable jobs and income. They left behind parents and grandparents and the memories of a childhood by the shore.[82]

By the late nineteenth century former Cape Codders were organizing trips to get their families back to the Cape for summer visits. The more successful built

The Decline of the Established Economy 103

summer homes near their old homesteads while others moved in with relatives or made arrangements to board with a Cape family. Florence Baker remembered her childhood when "city relatives" came by train to South Yarmouth, where they loaded their luggage on the stage and rambled along over the sandy roads to Bass River.[83] She noted an early vacationer to Bass River, a Mrs. Hughes who, according to family legend, was so enthralled by Thoreau's description of the Cape that she packed up her two daughters and took the train and stage to Bass River. There, she boarded with Mrs. White for ten dollars a week, five for her and two-fifty for each daughter.[84]

Mrs. Hughes and her daughters boarded with Mrs. White because she did not have relatives with whom to stay. As the summer traffic increased Cape Codders began to specialize in taking in visitors during the summer.[85] With the youth of Cape Cod moving on to mainland cities, they left behind older family members, many of whom were unable to move, unwilling to seek new jobs, or simply could not find employment in their traditional crafts. With the younger generation "remove[d] to the large cities" rooms opened up and taking in a summer boarder brought in necessary extra income. The Falmouth whaling captain Leonard Dexter built a comfortable home to which he planned to retire after a successful career killing whales. He did retire, but unfortunately he died soon afterward, in 1853, shortly after his retirement. His wife, looking for income, converted the family home into a boarding house and then a postwar "summer hotel." By the 1880s the hotel was such a success that it was extensively enlarged and renamed the Dexter House Hotel.[86]

Although Florence Baker remembered most visitors bunking in with relatives or taking a room in a boarding house, more and more visitors to the Cape, came, not as relatives, but as vacationers interested in the scenic beauty and fresh clean air of the seaside. The thriftier of these visitors looked to board with a Cape family. But as word spread that Cape Cod's beauty rivaled or surpassed that of the traditional summer retreats of well-to-do tourists, more city folk began to search out summer accommodations on the Cape. When Thoreau hiked across Cape Cod he noted only a few guesthouses for visitors, the sparse Pilgrim House being one. Yet within a little more than a decade, there were dozens of guesthouses on the Cape, one of the largest of which, at Cotuit Point, could accommodate several hundred. Truro, which had seemed a vast wasteland to Thoreau, expanded an original stage stop house into a large guesthouse on the Highlands.

With the coming of the railroad more people looked to the possibility of opening large "summer hotels" that would cater to the urban upper classes. The old Iyanough House, built in 1832 to accommodate people taking the packets to Nantucket, expanded and refinished itself in the 1880s to become a fashionable hotel for "summer sojourners" and take advantage of the extension of the Hyannis branch of the Old Colony Railroad.[87] By then Nathan Crosby's family had transformed Albert Crosby's dream retirement estate, Tawasentha, into a luxurious tourist hotel. With the arrival of the train came the dream of money to be made supplying the needs and wants of the off-Cape wealthy.

From the vantage point of the 1930s, L. Stanford Altpeter recalled that during the preceding fifty years, a strange new industry had developed on the Cape: "feeding, lodging, clothing, transporting and entertaining summer visitors."[88] The new summer hotels offered the "American plan," in which the patron paid a set fee that included room, board, and entertainment. Individuals or families from Boston or New York could take a train to any Cape Cod town they chose to be picked up by a coach sent from the hotel. In Cotuit the Santuit House, which began taking in guests in the 1860s, greatly expanded in the late 1880s to accommodate a hundred guests. The Crosby family of Osterville diversified their boatbuilding activities by building the Crosby house for summer visitors in 1876. Usually overlooking the sea, the new hotels offered croquet and tennis courts on their grounds, gardens for strolling, and often rowboats or small sailboats for the enjoyment of their more energetic guests. Most of the large summer hotels were located on the south shore from Falmouth to Chatham and were patronized, according to an 1890 observer, by "pleasure seekers" who sought out the warmer and more manageable water for swimming and recreational boating. On the north side of the Cape, where the water was cold, the Nobscussett House, located on an old whaling operation in Dennis, covered 125 acres and had almost a mile of seafront for bathing, boating, fishing, rambling, croquet, lawn tennis and swings. To feed its many guests the Nobscussett House arranged for the neighboring Tobey Farm to supply fresh milk and cream and for other other local farmers to supply vegetables. Fishermen brought in clams, lobsters, and fish.[89] In the evenings the hotel organized clambakes. Many of the hotels or large inns like Baker's Chequessett Inn added cottages to their grounds to cater to guests wanting to stay for longer periods of time.

Hotels and Vacation Colonies

Although most nineteenth-century Cape visitors came to the summer hotels or homes taking in boarders, some recreation entrepreneurs, including

The Decline of the Established Economy

the tourist visionary Lorenzo Dow Baker, were experimenting with new means of capturing the attention of the urban visitor. In 1889 Sheldon Bell bought up land in Truro on the back side at Ballston Beach and created the Ballston Beach Colony. Bell built a clubhouse, a dining hall, and several small rustic cottages. Local farmers who were more than willing to focus their production on vegetables and fruits, chickens, or sometimes dairy cows supplied food for the colony. Bell advertised his colony in the New York papers and looked to attract city people interested in a simple but comfortable vacation on the beach. In 1898 Baker bought land on the Truro bay side and built Corn Hill Beach Colony, a cluster of small cottages to be let out to families who wanted the privacy of their own cottages but could not afford either to own a summer home or stay at the more expensive hotels. In Craigsville, developers followed the model of the old Methodist camp meetings and developed a secular campground specifically for "the less well off summer visitors," which drew thousands of guests as early as the late 1880s.[90]

Boston visitors could take the train directly from South Station while New Yorkers would take the overnight ferry to Fall River and then catch the morning train to Middleboro, where they transferred to the Old Colony Rail Road. Once on the Cape, visitors confronted the problem of moving luggage and other summer paraphernalia to the hotel or guest home. For those who were visiting relatives, the family wagon or carriage was enlisted to get from the depot to the comforting cup of tea or lemonade. The larger hotels ran regular coach service between the train station and the hotel. Those staying in guesthouses or the smaller hotels depended upon local entrepreneurs who enlisted the family horse to do summer duty by carrying visitors from and back to the depot.

Soon "summer hotels" sprang up in all the towns across the Cape. But although more and more people were coming to the Cape, the recreation industry remained in its infancy. The Massachusetts Department of Labor and Industries noted in the 1870s: "Only a comparatively small number of persons seeking rest and recreation had discovered the attractiveness of Cape Cod for summer residence."[91] The 1890 *History of Barnstable County*, acknowledging that tourism was becoming important in Falmouth and Hyannis, failed to note any significant vacation activity in other places, even Provincetown.[92] The Massachusetts Bureau of Statistics of Labor responded to the dramatic decline of population on the Cape, and the concern that it had become an impoverished section of the state, with an investigation and a report in 1896. Although it noted that summer visitors were increasing on the Cape, the report dismissed

With the expansion of tourism, more and more Cape Codders began to rent out rooms or cottages to visitors from off-Cape. Photo courtesy of the Library of Congress.

the idea of their being a solution for the region's depressed state. It recommended instead improved truck farming, better freight rates, more irrigation, and land reclamation projects.[93]

Although tourism was still new to the Cape, in the late 1880s and 1890s hotel and inn construction increased dramatically. The *Chatham Monitor* noted in 1875: "Strangers and visitors are beginning to appear in considerable number. . . . Some require more and more rooms and privileges than can be given by a private family. It would be well if someone who had money to invest would build and furnish a few cottages to rent for those who would be glad to occupy them during the summer months."[94]

However, even the most elaborate and well-positioned accommodation on the perfect spot was no guarantee of success. Taking note of the concern of the *Chatham Monitor* about a lack of accommodations for the "strangers and visitors" who were appearing in "considerable number," two Boston businessmen, Eben Jordan (of Jordan and Marsh) and Marcellus Eldredge, bought up Nickerson Neck, a piece of mostly vacant land that stretched out into Pleasant Bay, for five thousand dollars. They also bought Strong Island in Chatham, which had previously held a few fishermen's homes and an abandoned brickyard. Eldredge was born and raised in Chatham but moved to Portsmouth, New Hampshire, where he eventually came to operate the largest brewery in New England. He came back to Chatham and bought a summer home on Watch Hill. Eldredge and Jordan proceeded to build a huge luxury hotel, which opened its doors in 1890, and capitalized on the Cape's natural beauty and its quaintness: "Born of generations of mariners and in the very spray of the ocean wave, the inhabitants of Chatham are through and through a seafaring people . . . a modest and amiable race." The hotel also advertised itself as built "upon the very brim of the sea. The hotel property [has] . . . between 6 and 7 miles of continuous water front."[95] The hotel had parlors, a billiard room, dining rooms, seventy bedrooms with water views, and the latest plumbing. It offered bathing, gunning, fishing, archery, horseback riding, and clambakes. But with all that going for it, the hotel failed after three seasons.[96] The *Chatham Monitor* may have felt there was a need for more accommodations, but local residents opening up their homes as boarding houses provided too much competition for the luxurious hotel that opened just a decade or so too soon.

The Chatham Inn was not the only project that mistimed the Cape's switch from the regime of production and extraction to that of tourism and vacationing. When salt making and boatbuilding declined in Hyannis Port, on the south shore along the calm warm waters of Nantucket Sound, the price of shorefront land dropped. Hoping to take advantage of cheap land and an anticipated interest in beachfront property, the Hyannis Port Land Company bought up the land for a resort development. Unfortunately, the land company was too farsighted. The development failed and by 1890 the land fell into the hands of a Framingham, Massachusetts, bank.[97]

Cabins to Cottages

The decline of agriculture, salt making, and boatbuilding may have been hard on Cape families, but it did not hurt the hunting and fishing, and it did reduce the price of land. For those with a little extra cash, it was possible to buy some property along a good fishing pond, hunting area, or shoreline. Initially, simple fishing or hunting shacks were built, but as the century ended, families like the Underwoods began to look to the Cape as more than a place to shoot an animal or hook a fish; these simple shacks were soon upgraded to summer vacation homes.[98]

Upgraded hunting and fishing shacks were not the only source of summer homes on the Cape. As Ezra George Perry well understood, money was made in the city, but living in the city during the hot, polluted summer months was hard. Money offered an alternative. For some, that alternative was a weekend, a week, or a season at one of the fashionable hotels rising up across rural New England. Others looked for a more permanent solution: a place where the family could go to escape the heat, crowds, and pollution of the summer; a place away from demands of the job and away from the diverse, increasingly foreign-born populations that were pushing into the urban centers of the Northeast; a private place for family, friends, and ethnically similar people. The exclusive hotels offered some of that, but not the privacy nor the guarantee of similarity that could be bought by owning one's own estate.

The Buzzards Bay area of Falmouth and Bourne was the first place in which the wealthy began to construct large summer estates. In 1852 Joseph Fay, a wealthy Bostonian, chanced upon an impressively scenic stretch of farmland in Woods Hole, overlooking the harbor. Local farmers desperate for any income gladly sold their property cheaply to Fay. He then built an elaborate summer home.[99] Soon others followed. In 1892 Charles Jones, a developer from Wellesley, chanced upon a thirty-acre peninsula in West Falmouth cluttered with ramshackle wharves and discarded fishing paraphernalia. The owners were more than willing to sell to Jones for eight hundred dollars. The Beebe family, three brothers who were successful Boston businessmen, bought a number of abandoned farms between Falmouth Village and the shore. Wanting to create an "unspoiled rural atmosphere" the Beebes brought up land around their initial estate whenever they could. They also encouraged their wealthy friends to join them. Eventually the area took on the "resort characteristics of manorial exclusiveness."[100]

By 1910 Falmouth had become a prime location for elegant summer hotels and summer estates. Edmund Davis gave up making salt in the 1870s and sold off the land in Falmouth Heights to a group of wealthy Worcester businessmen who sold lots for summer homes and began buying more land; word spread in Worcester that Falmouth Heights was the fashionable place to summer.[101] Falmouth Heights soon became a prime location for summer homes for the wealthy of Worcester. Likewise, a group of Brockton shoe manufacturers bought up land at Silver Beach and established a summer colony there. In 1880 the Keith family of Newton built a substantial summer cottage at Megansett Beach, and soon Newton friends were buying surrounding chunks of land. Word spread also to Watertown, and the wealthy of that town colonized neighboring Megansett Shores.[102] By the 1880s Falmouth's old identification with the products of the sea was a fading memory:

The Decline of the Established Economy 109

the "prominence of this vicinity as a summer resort . . . [was] steadily increasing and rapidly becoming the chief characteristic of the town."[103]

As Ezra George Perry traveled about the Cape at the turn of the century he reminded his readers that the Cape has become a place where "Boston's business men, who first came from the city for rest and reviving their health [were] now building homes to summer more permanently."[104] And the price to get in was increasing. Land developers like Perry were buying up land as fast as they could. Perry noted that his father had sold an island in Buzzards Bay for two hundred dollars that was now worth over seventy-five thousand. Land on Monument Neck that sold for four dollars an acre "a few years ago" had so increased in value by 1898 that "money [could] not buy it."[105] The land that local Cape Codders were selling was mostly land used under the older regime of extraction and production. By the second half of the nineteenth century this activity had ceased or was greatly reduced. Harbor inlets used as whaling centers, ship works, or saltworks had been abandoned and were now for sale at low prices. The wealthy bought up huge swaths of this land and then sold off chunks of it to their friends.

As more elegant summer estates were built and land prices began to climb, developers such as Perry rushed in to purchase land from locals before the prices began to rise.[106] As early at 1882 a map of Cape Cod included an advertisement offering for sale "a very attractive and valuable piece of Sea Shore land . . . with great frontage on Buzzards Bay. Handsomely wooded . . . *a rare chance for speculation.*"[107] This land had not even been laid out for development, but within ten years of the publication of the map it was laid out and sold. Charles Jones took his eight hundred dollar piece of land and hired the famous landscape architect Ernest Bowditch to have it cleared and landscaped. The plan called for generous lots and the property, renamed Chapoquoit Island, sold for fifty thousand dollars. Similarly a farmer who believed an old sheep pasture on Great Hill to be worthless sold it for almost nothing to developers who laid out lots and transformed it into one of the town's most prestigious areas. Even the old guano and fish oil fertilizer plant on Penzance Point became an exclusive development after the decline of local fish led to its abandonment as a fertilizer plant.[108]

Perry's vision was of a Cape with "elegant mansions" for "some of Boston, Brooklyn and Chicago's richest men."[109] Among the mansions built on the Cape in the late 1880s were President Grover Cleveland's Gray Gables and the summer residence of the period's most famous actor, Joseph Jefferson. For Perry, the old Cape Cod of fishing and farming was not the golden age; rather, it was the "Cape Cod of today" epitomized by Falmouth, "the Golden Garden City, where the old hills used to sell for $5.00 per acre and the new hills now [end of the century] sell for $2,000 per acre, all this inside of 20 years."[110] The reason land was selling for five dollars an acre in 1880 was that the old Falmouth had collapsed. The whaling fleet was sold or was rotting on sand bars. The fishing fleet was shrinking. Saltworks such as Edmund Davis's had been abandoned.

Falmouth was only one of the towns where the wealthy established comfortable summer estates on the water. The stagecoach service between the Old Colony Line and Wianno Beach in Osterville, where the Crosby family had already established a resort hotel, encouraged developers to buy old saltworks and convert the land to an expensive summer resort.[111] Abandoned saltworks and shipyards across the Lower Cape were being bought up and then sold off as lots to the wealthy. Summer homes and estates came to dominate the shoreline, not only in Falmouth and Osterville but also in Centerville and Hyannis Port. In 1838, Wellfleet's old 1765 mill was moved from Mill Hill to a spot overlooking Wellfleet Harbor to take advantage of winds blowing up from the sea. By 1850 it discontinued grinding grain. At the end of the century Mrs. Heller, a wealthy Bostonian, bought the mill and turned it into her summer home, known locally as Morning Glory.[112] The fate of the old mill in many ways reflects the fate of the Cape itself.

Henry David Thoreau appreciated the rugged beauty of Cape Cod. He believed that eventually its beauty would lure more and more visitors to its windswept vistas. He did not believe that its beauty would ever be enough to make the Cape a fashionable watering hole for the wealthy. Thoreau was a very perceptive observer of nature who believed the Cape would never become a fashionable resort community, but Ezra George Perry was out to prove Thoreau's prediction wrong.

III
DEPENDENCE ON DISTANT RESOURCES, AND REVENUE FROM RECREATION
Early Twentieth Century to the Present

6
TRAINS, CARS, COTTAGES, AND RESTAURANTS

In the hundred years or so from the end of the Cape's golden age until the end of World War II, Cape Cod slowly lost its long-standing fishing, farming, and forest economy and replaced it with all the activities that cater to the needs of tourists, summer residents, and retirees. Even those elements of the older productive economy that persisted, came to depend more and more on tourism for their survival. The new reality built upon nineteenth-century precedents and visions: Lorenzo Dow Baker's imaginings that camp meetings could become cottage settlements for middle-class families; Ezra George Perry's vision of abandoned farms sprouting estates for the wealthy; Loring Underwood's conversion of his hunting lodge into a summer family cottage; and the Crosbys' longing to return home to the family place of origin.

These projections came to be realized in two distinct eras. The first era, 1895–1920, continued the weary process of resource exhaustion that characterized the Cape's failing economy of fishing, forestry, and farming. The arrival during these same years of wealthy estate owners in the Bourne/Falmouth area and the success of the "American plan" hotels prefigured the later vacation Cape.

The second era, 1920–1950, grew with the new mobility of the automobile, the improvements of roads, and the arrival of electricity. A boom in inexpensive cottages on inland sites, or on bits of abandoned shore property, reversed the long trend of population loss. Federal government public works coupled with the construction of state highways supported the transformation of the Cape into a popular middle-class resort serviced by local residents and a seasonal army of young workers. By 1950 the fashions of the rich (their yacht clubs, golf and tennis clubs, sport fishing, and camps for children) extended into the rhythms of middle-class family life on vacation.

BARNSTABLE COUNTY POPULATION

1850	35,276
1860	35,990
1870	32,774
1880	31,897
1890	29,172
1900	27,172
1910	27,542
1920	26,670
1930	32,305
1940	37,295
1950	46,805

Despite the commonalities of these resort patterns, Cape Cod in 1950 remained a place of distinct towns and special local traditions. Out of this remnant, a powerful mythos of the "old days" was constructed as an alternative to the commercialism of real estate promotion and retail exploitation.

The fifty years from 1900 to 1950 continued to follow the Cape's ever-changing environmental processes: everything was in motion and nothing remained fixed. The population, the visitors, the fish, the farming, the housing, and the transportation all changed, each in its own way, together forming a totally new Cape environment. Nevertheless, as it has from the start, the Cape ecosystem remained fragile.

FARMING FOR LOCAL MARKETS

Although it took a generation before tourists came to dominate the Cape's economy, their growing presence affected the Cape's older industries. Cape farmland had been ill used during the first half of the nineteenth century: traditional agriculture had declined and sheep and beef cattle raising had collapsed. Grass, bushes, and trees now settled in on abandoned fields to hold the soil. With the exception of cranberry production, which expanded to occupy more than three thousand acres and produced over six hundred thousand dollars' worth of berries in 1890, agricultural activity suffered through most of the late nineteenth century. Commonwealth investigators noted in 1922 that, "Cape Cod has been allowed to grow up huge tracts of wild brush land [allowing the soil to be] enriched by leaf mold for a long period."[1] Farmers, facing steep competition from the American farm belt, gave up attempting to produce their market share of cereal crops, beef cattle, and sheep and returned to the practice of their great-great grandparents. Once again they started using marsh hay, animal manure, and green cover crops to restore fields. They also concentrated on crops that favored acidic sandy soil.[2] Turnips, beets, onions, potatoes, asparagus, strawberries, apples, and blueberries spread out across the Cape from Falmouth to Truro, joining the barrels of cranberries heading for Boston markets.[3]

Despite this increase of marketable produce, the state's investigation of the depressed conditions on Cape Cod argued that farming "cannot be made profitable . . . until there are more favorable freight rates and better facilities for reaching city markets."[4] What the state investigators did not take into account was the significance of farm stands and the growing local demand for "market gardening."[5] Although Eastham led the nation in asparagus production, local markets catering to the growing numbers of summer guests were what pushed the renewed development of Cape agriculture.[6] Ezra Perry noted at the end of the nineteenth century that in one successful hotel, the Chapman House, "the table [was] set with everything from their own farm."[7] Katharine Dos Passos and Edith Shay, recalling the early-twentieth-century Cape, remembered how farmland, once abandoned and windblown, was reclaimed to raise potatoes, melons, tomatoes, carrots, turnips, and other fruits and vegetables for local markets.[8]

Falmouth, though in decline as a center for fishing and whaling since the 1850s, found new wealth in its sandy soil in the early twentieth century. In 1890 Falmouth had a small Portuguese population who began in the whaling industry but looked for other ways to support themselves. By the turn of the century they had switched to farming, especially strawberry growing. Buying up abandoned farmland, these farmers put in strawberry patches and called on Portuguese friends and family to help with the harvest.[9] The combination of cheap land, tight family and community connections, and hard work paid off as Cape strawberries were in demand both on and off the Cape.[10] By 1920 Falmouth counted as one of the largest strawberry-producing areas in the country.[11]

The hotels, guesthouses, and inns also looked for fresh poultry and eggs. Cape farmers had always had chickens, but the expansion of the summer tourist trade created a local market for both eggs and chicken meat. Chicks hatched over the winter began laying just in time for the summer trade. With the coming of winter and the molting season, laying declined as did the demand for eggs. When the laying cycle began again with new pullets, older hens were sold for stewing or roasting. The expanded demand for poultry led to the creation in 1921 of the Cape Cod Poultry Organization.

Local markets also encouraged farmers to get back to the neglected orchards and open up fields for tomatoes, lettuce, and melons.[12] The state investigator looking into the late-nineteenth-century depression on Cape Cod noted that although tourism was not a major factor in the Cape's economy, it did help truck-farming operations that catered to "the ever-increasing demand in the summer months."[13] And Cape farmers organized and diversified.[14] In 1916 the Cape Cod Farm Bureau emerged to encourage and market Cape agricultural products.

Farmers who were refurbishing their farms to meet the growing demands of the tourist hotels added roadside stands to their operations to sell fruits, vegetables, chickens, ducks, and eggs.[15] In 1920 Cape farmers worked more than forty-seven thousand acres of active farmland.[16] By that same year the Cape was producing nearly a million dollars' worth of agricultural goods, with fruits accounting for well over half. More than 18 percent of the Cape's land was being farmed.[17] Most farms were small affairs of approximately ten to twenty acres with farmers harvesting a third to a half of it.[18] The most successful farmers sold at roadside stands.[19]

However, even with the increased demand for fruits and vegetables, by the 1920s most Cape farmers were in fact part-time farmers. Anthony Marshall, a Truro resident, remembered how his father covered both sides of the summer demand. He not only tended the family farm, which comprised several acres of land in turnips, parsnips, carrots, potatoes, corn, broccoli, asparagus, fruit trees, chickens, a couple of pigs, and a cow, but he also worked as a builder of new vacation homes.[20]

FISHING IN "SCARCE WATERS"

Across the Cape the decline in fishing continued into the twentieth century. Weir, or pound net, fishing increased in the late nineteenth century, providing fish for local fish processing plants and for markets off-Cape as well as bait for cod boats. But by the late 1890s, the numbers of these fish declined.[21] Mackerel practically disappeared from the coast at the end of the century. There was considerable debate among fishermen over the cause of this decline. Offshore fishermen blamed the pound nets set along the shore, whereas the lower-capitalized shore fishermen argued that their nets did not undermine the fisheries.[22] Low prices and a "scarcity of fish" hurt the cod fisheries, and the more heavily capitalized vessels undercut the smaller fishermen. The emergence of the large beam trawlers transformed cod fishing in the twentieth century even more than long-lining had in the nineteenth century. Beam trawlers utilized the same dragnetting techniques as the ground fishermen did. The process damaged spawning grounds and injured young fish, and it required more power to work the nets. This ended the era of the schooners. A beam trawler with a crew of eight could haul in loads of 250,000 to 300,000 pounds of fish in one outing; a schooner with a crew of twenty-four working long lines would be lucky to bring in 70,000.[23] Nor could the larger vessels negotiate the shallower harbors of most Cape towns. The new beam trawlers quickly dragged clean the near-shore shoal grounds, forcing the fishing fleet farther out into deeper water.

Although the number of fishing fleets heading out from Truro to Falmouth (and towns in-between) declined in the late nineteenth and early twentieth centuries, Provincetown fishermen persisted. During the early years of the twentieth century Provincetown's fishing fleet's catch placed just behind those of Boston and Gloucester. It did so by fishing the Georges and Grand Banks.[24] The continued success of Provincetown's fishing fleet allowed the town to hold its population steady while other Cape communities saw dwindling of their populations. A good many of the Provincetown residents were Portuguese fishermen, descended from families who had come to the Cape as whalers. When the whaling industry declined, they shifted to cod fishing and encouraged their relatives across the Atlantic to join them. By the 1890s the Portuguese made up almost 25 percent of Provincetown.

Fishermen, who had been frustrated with market prices and smaller hauls for their traditional catch of herring, cod, and mackerel, now found that tourists preferred local fresh fish.[25] Groundfish, flounder and halibut, swordfish, and fresh cod joined oysters, scallops, clams, and lobsters on the menus of hotels, inns, and restaurants across the Cape. The increased demand for fresh groundfish had an unfortunate outcome. A new fishing technique, whereby the fishermen lowered dragnets to the ocean floor to scoop up anything they could, caused a crash in the flounder population in the early 1920s around the south coast of the Cape and across all of Cape Cod Bay. The state attempted to limit flounder dredging between May and November, but smaller catches led to higher prices. The new prices, along with the pressure of more expensive equipment, led to significant flouting of the law.[26] The flounder dragnets not only swept in massive amounts of groundfish, but they also disrupted the feeding grounds and spawn of cod.

Oyster fishing also began to suffer from overfishing and other destructive practices, just as the demand for oysters began to climb in Wellfleet. Overharvesting of oysters in the late nineteenth and early twentieth century, along with the removal of shells from the water to use as calcium for road cover and liming fields, cut yields of the town's oyster beds dramatically.[27] Only later in the second half of the twentieth century with the development of a closely monitored grants system, did commercial oystering slowly recover.

BOATBUILDING, PARTIALLY RECOVERED FOR SPORT

Farmers and fishermen hung on by shifting to the emerging tourist market and by sending their wives and daughters out as maids and cooks. Boat builders also shifted to the tourist trade. Captains of old fishing boats may or may not have transformed themselves into "skilled and careful skippers" for hire by sport fishermen as Perry suggested, but Cape waters, particularly those of Nantucket Sound and Buzzards Bay, gained a reputation for sport sailing.[28]

The boat builders who had managed to hang on while orders for traditional fishing boats declined now found new customers interested in vessels for sport sailing. The first three decades of the twentieth century were the golden age of American yachting—and American yacht maintenance. Having a prestigious Herreshoff-built yacht delivered to a mooring off one's property may have been the mark of class and privilege, but once built and in the water even these well-designed and skillfully built boats demanded constant maintenance. In winter their hulls needed to be scraped down and repainted. Their metal fasteners, in contact with salt water, caused iron sickness in wood, which required constant mending to maintain the integrity of the craft. Rudders and keels regularly needed repair or replacement. Local boat works across Buzzards Bay, Hyannis Harbor, and Pleasant Bay found new work repairing and maintaining these new vessels that increasingly filled their harbors. The old Crosby catboat came back in style as a sailing craft for the younger generation of wealthy vacationers.[29] The Chequesset Inn in Wellfleet was one of several inns to hire local boat builders to fashion a large number of skiffs, which were made available to guests for rowing about.[30]

LAND FOR THE WEALTHY

The maintenance of the Cape's traditional activity depended upon visitors with money coming to the Cape, which in turn required marketing the Cape as a vacationland to those who could afford vacations. Boston's wealthy class had long been of the habit of leaving the city during the hot summer months for the cleaner and presumed healthier air of the country. Boston's elites established their summer homes and farms along the tree-lined streets of the outer reaches of the city of Boston and its inner suburbs, such as Chestnut Hill, Brookline, Newton, and Belmont. As the city and its transportation system expanded into what had been the rural countryside of the early nineteenth century, well-to-do owners of these summer homes and farms converted them into permanent residences. But the longing to escape the city, even its outer edges, remained. For the upper middle and professional class, escape often meant a week in the country staying at an inn, resort hotel, or country farm.[31] Ezra Perry realized that these retreats were not exclusive enough for the upper classes. They were not the social enclaves of their suburban settlements. He also realized that the wealthy not only looked for exclusivity, but they wanted to have their retreats embedded in enduring property, something that could be passed on to future generations.

The interest of the wealthy in owning coastal estates, removed from the city and its increasingly foreign population, was certainly not something Ezra Perry invented. Newport, Long Island, Marblehead, and Martha's Vineyard were all well established as watering holes of the gilded class. Perry's contribution was to focus the attention of the wealthy on Cape Cod. His was not an easy task. In an 1875 guidebook to New England, Samuel Drake described the Cape as a place of "widespread dissoluteness." Appleton's *Illustrated Handbook of American Summer Resorts* the following year dismissed the Cape with a mere one-page description, calling it a "wilderness of sand."[32] Fourteen years later the Old Colony Railroad admitted that the Cape was a "treeless, almost verdure-less barren plain, windswept and bleak."[33] Although the railroad offered visitors easy access to the Cape, this was not its main source of revenue. The railroad made more of its money shipping fish, oysters, cranberries, asparagus, and strawberries to inland urban markets. The 1890 history of Barnstable County that mostly celebrated the wonders of the Cape noted some tourism in Falmouth and along the south coast to Dennis. But in discussing towns further out on the Cape—Harwich, Brewster, Orleans, Chatham, Eastham, Wellfleet, Truro, and Provincetown—the 1890 history touted fishing and agriculture as important parts of the local economy, with no mention of tourism.[34] A history of the Old Colony Railroad in the 1910s noted that in the late nineteenth century, Cape Codders lived in tight communities hostile to visitors. Smelly fish processing still dominated many of the beaches and shorelines, encouraging flies but not tourists.[35] Pound nets dominated shore vistas while the smell and noise associated with harvesting of the fish caught in these nets undermined the ideal of a tranquil coastal escape. Although by 1873 Falmouth had good train linkages to Boston and New York, a foul-smelling plant converting guano (also known as bird droppings) and ground-up fish to fertilizer overwhelmed one of the town's most picturesque beaches.

Despite these drawbacks Falmouth was ahead of most of its sister Cape towns in looking to summer visitors. The Fays, who came into town in the midcentury, bought up cheap abandoned farmland around Woods Hole, and the Beebe family bought up farmland around Falmouth village. The Beebes and the Fays built huge estates and encouraged their wealthy friends to do so likewise. By the 1880s, sections of Falmouth took on the "characteristics of manorial exclusiveness," and an enterprising developer in 1882 produced a map of the area advertising "for sale a very attractive and valuable piece of sea shore land . . . with great frontage on Buzzards Bay . . . handsomely wooded . . . a rare chance for speculation."[36] By the mid-1890s the land was laid out and sold.[37] But these islands of exclusive estates were not the norm even in Falmouth. Other areas of the town were still occupied by the remnants of the older activities. And one person's "handsomely wooded" property was another's mosquito-infested, brambled wilderness.[38] Although the fishing industry was in decline in the late nineteenth and early decades of the twentieth centuries, working boats and commercial marine activity still filled the Cape's many harbors and wharves, leaving little room or patience for yachtsmen and pleasure boaters.[39]

Ezra George Perry looked beyond the decaying fish waste floating about along the shore and the jumble of the working harbors to a different and future Cape Cod, one of clean beaches, crisp fresh air, and moorings filled with yachts and pleasure boats. He stated in his 1898 *A Trip around Cape Cod: Our Summer Land and Memories of My Childhood,* "We are living on the new Cape Cod or perhaps Cape Cod is destined to be one of the greatest watering-places this globe has ever seen. We have only begun in our advancement."[40] Fortunately for Perry, with each passing decade after 1860, there were fewer and fewer residents of those tight little hostile-to-outsiders villages, and the beaches and harbors were emptying themselves of their odoriferous activities. The saltworks that had lined the shore from Buzzards Bay to Provincetown were gone by 1890. With the dramatic decline of fishing came the collapse of fish processing; visitors to the Cape might still find a few fish-processing operations in Provincetown and Truro in the early twentieth century, but for most of the rest of the Cape they survived only in the memories of the old timers.

Fishermen and fish processors alike looked elsewhere for employment leaving behind "many empty houses." The Cape had about six thousand fewer inhabitants in 1890 than it had thirty years earlier on the eve of the Civil War; in 1860 almost thirty-six thousand people had lived and worked on Cape Cod. (The Cape would continue to lose population until 1920, when only 26,670 called the peninsula home. It was not until 1940 that the Cape recaptured the population it lost over the sixty years after the election of Lincoln.) So as the nineteenth century drew to a close, with fewer and fewer boats leaving to fish the banks, or even tend to the pound nets that still lined the coast, old wharves were falling down into picturesque ruins while big heavy fishing schooners and banged-up working catboats were abandoning moorings and opening up space and a new esthetic.[41] In the mid-1890s a New York syndicate hoping to take advantage of empty homes and abandoned fish-works in Truro began buying up shorefront land for a summer resort.[42]

Perry's self-appointed task was to convert the public's image of the Cape—from a backwater of mosquitoes, drying fish, flies, barren sands, and hostile locals into a place of quaint villages, beautiful vistas, and even more beautiful estates peopled by the right sort. To do so, Perry was able to trade on the presence of wealthy pioneers who had established summer homes ahead of the tourist boom of the twentieth century. As he informed his turn-of-the-century readers, "There are more noted men located on Buzzards Bay shores than in all the rest of New England."[43]

As Perry understood, the Beebes and Fays were not the only wealthy pioneers enticing others to join them. To the east of Perry's home, the renowned naturalist Charles Cory bought all of Great Island off of Hyannis in 1883. Where previously fishermen and salt makers had worked, Cory built a summer home to view birds and hunt ducks. Hunting season brought Cory's wealthy friends from Boston and New York to the island for sport and leisure; among these visitors was Grover Cleveland, who soon built his own summer estate on Buzzards Bay.[44] This changed Cape, dotted with estates of prominent New York and Boston families, was the Cape that dominated all of Perry's books, especially *Trip around Cape Cod*.

But Perry was first and foremost a salesman. He was busy buying up cheap Cape land and offering it for sale to the wealthy for building sites for summer estates. He also offered his services selling land for others and renting out already-built homes. He meant his travel books to advance his real estate interests. In an attempt to disabuse future clients of the notion that the Cape was an isolated wilderness of hostile villagers, his books were packed with pictures of wealthy estates and coastal sport-boating communities, stately churches (particularly Protestant churches), and tree-lined streets with libraries and neat country stores. The descriptions of the towns focused on available amenities and local residents who were friendly, helpful, and cheerful.[45] Readers did not find accounts of smelly fishing piers or see pictures of decaying fishing boats or fish processing plants in Perry's books. Farms were "well kept," producing "beautiful crops." Although he chose *Our Summer Land and Memories of My Childhood* as the subtitle for *A Trip around Cape Cod*, the Cape he spent most of his time describing was not the Cape of his childhood but the Cape of a "wonderful change for the better."[46]

In that book Perry took his readers on a tour, through quaint villages peopled with hearty native-born residents, to showcase the Cape's great potential for new real estate investment. Perry's Brewster had "elegant summer homes of rich and retired people," including the home of Roland Nickerson who, Perry noted, made his fortune in Chicago but came back to the family homestead to build a "beautiful" retirement home. Although Wellfleet still had a functioning fishing fleet and a significant oyster industry, Perry described the town as "getting to be quite the summer place." Perry's Harwich was a community of "many very handsome summer cottages . . . built within a few years"; Hyannis was "going to be one of Cape Cod's greatest summer places."[47]

Perry countered the image held by many New York and Boston residents of Cape Cod as a smelly, isolated backwater by projecting the Cape as healthy and beautiful with "cool and refreshing sea-breezes." He claimed that successful Bostonians would be surprised that already many Cape towns had become "very fashionable" with "wealthy people" as summer residents.[48] Even the old Cape served the new." For those who love to sail and troll, a fleet of white-winged fishing boats under skilled and careful skippers within hail will soon bear [one] out in deep water amid a lively school where you can pull in the great mottled beauties hand over hand . . . right in sight of your island home."[49] These sport-fishing boats were not the smelly beat-up crafts of old, piloted by curmudgeon old salts, but lovely, clean boats captained by "skilled and careful skippers." Perry's new Cape was, indeed, a place where a "great change has been made" and it was "a wonderful change for the better."[50]

That change not only meant that stores would now "run according to city methods and [carry] a full stock of fresh meats, groceries and provisions," but it also heralded the coming of possibility.[51] The new Cape Cod was a place of opportunity for those who bought in early. "Do you want to invest some money that will bring you big returns?" Perry asked his readers. "There is a great deal of money made in buying land on Cape Cod."[52] Perry focused on the value one could get by buying land while prices continued to rise. "Cow pasture . . . at one time only valued at about five dollars per acre," was sold off "at a high price" after being divided into lots. The land on Monument Neck in Bourne (just south of Grover Cleveland's estate at Gray Gables) sold for four dollars an acre "a few years ago," but it had increased in value by the time Perry was writing, so much so that "money cannot buy it." Perry reminded his readers that locals were selling low, and developers and early buyers were benefiting. On Chocequoit Island, also part of Bourne, "five years ago there was not one house on it. An enterprising gentleman from Boston saw its very attractive location and invested money on the island and then began to interest his friends." The island, worth a few hundred dollars in 1890, had $150,000 worth of taxable property on it by 1898.[53] The rose-colored image Perry depicted provides a glimpse of the Cape he wanted and was partially instrumental in bringing about.[54]

Perry, who bought up land along Buzzards Bay, wanted to sell property, but his new Cape offered opportunity for others as well. As Perry pointed out, the railroad brought change to the Cape. Although its primary source of revenue in the nineteenth century was the movement of produce, the railroad also meant that travelers from Boston and New York could now reach the Cape in a day or a night and a day. For the wealthy who were not interested in owning their own estates—and for those without the extra financial resources to tie up in a summer home—large, luxurious hotels that had begun to develop in the nineteenth century came into their own. Although the large estates came to dominate Buzzards Bay, the land to the east was a mix of estates, hotels, and guesthouses.

HOTELS, GUESTHOUSES, AND UPSCALE AMENITIES

The luxurious Chatham Hotel failed in 1893 after three seasons—although others to follow it would succeed.[55] (The old Chatham Hotel was pulled down but not until 1922 was the land converted to the Chatham Country Club.[56]) Soon after the hotel closed other developers began to speculate on land along the Cape Cod Bay and the Cape's ponds. In the years before the World War I, Charles Hardy formed the Chatham Associates with a group of local businessmen. They began buying abandoned "sandy beach plum and bayberry land along the shore" for as much as five hundred dollars an acre, a price locals thought crazily high for abandoned land.[57] Hardy's syndicate then built the successful Chatham Bars Inn with an accompanying nine-hole golf course, and around Mill Pond he built substantial cottages for sale to the "rich and successful."[58]

Other towns saw the expansion of the grand luxury summer hotels on the Cape. Lorenzo Dow Baker's Chequessett Inn, in Wellfleet, offered "sea and lake fishing, boating, bathing, tennis, billiards, bowling and orchestra" with food from local farmers, fish from local fishermen, and a room overlooking the sea at a cost of fifteen dollars a week in 1916.[59] Across the Cape former sea captains' houses were enlarged to accommodate guests; guesthouses and hotels and inns expanded and new ones sprung up along both Nantucket Sound and Cape Cod Bay.

Getting to the Cape, even by railroad, involved much planning and coordination. Summer visitors told of spending a whole day gathering up and packing trunks, hiring a wagon to get the trunks to the station, and another day moving the trunks to the guest house or hotel and unpacking.[60] Once on the Cape, visitors' compass remained close to the hotel. It was possible to walk the beach, but one had little access to the larger Cape. For those Americans with significant discretionary money and leisure time, the resort hotel offered a complete package. A guest was housed, fed, and entertained in a beautiful setting, and usually, if the accommodations had been chosen carefully, surrounded by like-minded people of the same social class. The Highlands House in Truro, for example, offered room and board for anywhere between $2.50 and $6.50 a week and $1.50 a day before World War I.[61]

In the prewar years vacationers arriving on the Cape found accommodations in a variety of places. In addition to the large hotels and guesthouses, there were small cottages built by developers like Lorenzo Dow Baker and rented out for the week or month. The cottages varied from small one- to three-room shacks to "palatial estates" with eight or more bedrooms, drawing rooms, elaborate dining rooms, and servants' quarters, and were available in Chatham, Hyannis Port, and Craigsville. Most were located in or close to a town center and within walking distance of stores and the beach.[62] A small three-room cottage typically rented out for forty dollars a month; the hotels, whether the deluxe or more modest, operated on the American plan, in which room, board, and entertainment were included in the price, but cottage renters tended to fend for themselves.[63]

As Henry David Thoreau appreciated, Cape Cod was (and still is) a widely varied place. Thoreau described the Cape as a bent arm pushed out into the Atlantic Ocean. The shore facing the Atlantic on the Cape's forearm is rough and windy with wild waves crashing upon almost twenty miles of sandy beach. On the west of the forearm and north of the bicep, lay Cape Cod Bay, where the waters are calmer and shallower, particularly on the west side of the forearm. Nantucket Sound, to the south of the bicep portion of the Cape, like the bay, is shallow and calm. Inland from the shore there are 277 freshwater lakes and ponds in the midst of meadows and heathlands, rolling farmland, and cranberry bogs. The water temperatures in the Sound and along the bayside reaching north from Brewster to Provincetown are warm; the Atlantic waters and the bay waters from Sandwich to Brewster are cold. Scattered along the shore in the nineteenth century were quaint, worn, and run-down fishing communities; inland were small farming villages. Cape Cod offered hundreds of varied picturesque settings, for hiking and viewing along a strip of sand that stretched some sixty miles out into the Atlantic.

Yet most of this variety was not available to the hotel guest, whose experience of the Cape was confined to the area immediately surrounding the hotel or slightly further afield during a planned excursion. In that sense a Cape vacation was not a lot different from a vacation in Bar Harbor, Maine.[64] And if one were interested in a fashionable social life, a Cape vacation would probably be a disappointment compared to Newport with its rich array of yachts and yacht clubs, concentrations of wealthy estates, concerts, libraries, and tennis clubs. In the early years of the twentieth century, the Old Colony Rail Road offered special excursion runs during the summer weekends to encourage visitors to vacation on the Cape, and hotels and guesthouses offered a wide variety of activities and quality dining. In spite of these incentives, the Cape struggled in competition with other resort areas in the mountains of Vermont, New Hampshire, and Maine as well as the established resorts along the Atlantic coast from Long Island to Bar Harbor.[65] Cape resort hotels offered their guests recreational fishing, boating, lawn games, and clambakes. But so did the hotels in competing regions.

The Cape also had disadvantages aside from its limited social scene. The mosquitoes that so concerned Lorenzo Dow Baker did not plague only Wellfleet; early summer residents of Buzzards Bay also complained of mosquitoes.[66] And the Cape's cranky human residents were notorious for their unwelcoming ways. But despite the Cape's faults, the explosion in size and wealth of the American upper-middle and upper class brought visitors to the Cape in ever-increasing numbers and filled the growing numbers of guesthouses and hotels. Katharine Dos Passos and Edith Shay remembered that by the early years of the twentieth century "there were the rich and fashionable who swam, and yachted and fished and golfed in the villages of the upper Cape, Hyannis Port, Cotuit, Centerville, Wianno, Osterville, and others [particularly Falmouth]," while Dennis and Provincetown were gaining a reputation as centers for artists.[67] Dos Passos and Shay remembered the Cape of the early twentieth century as a place where "many families from all over the United States [made] the Cape their second home," where "they and their children [could lead] a life of fresh air, and fishing and boating and swimming."[68] Elroy Thompson noted in 1928 that increasingly Cape Codders were "tourist-minded or vacationland-conscious," with "the entertainment of tourists and summer residents bringing better returns than [fishing or farming]."[69] The old fishing and cranberry town of Harwich, after watching Chatham transform itself from a fishing village to a tourist center, cast about for a way to draw summer visitors to its shores. The land around Wychmere Pond, which bordered Nantucket Sound, had an abandoned track used in the nineteenth century by local young bucks to race horses. Harwich dredged a cut

Trains, Cars, Cottages, and Restaurants　　　127

between the pond and the Sound to create Wychmere Harbor for summer boaters and built a golf course to entice the wealthy to settle in the town. Although much of the shoreline to the west, bordering Dennis, remained undeveloped in the interwar period, the area around Wychmere Harbor saw the construction of substantial summer homes.[70]

PROMOTION: IMAGE AND IDEAL

The real estate development that so excited Ezra George Perry also engendered concern among some Cape Codders who worried that the shift to the tourist economy would make it difficult for the Cape to preserve what made it desirable in the first place. In 1901 a group of well-connected Cape Codders, mostly businessmen and bankers, established the Cape Cod Improvement Association. The association recognized that the Cape's future "of necessity . . . will be different from the past" and that "it is fair to assume that the only future for Cape towns lies in developing them as summer resorts," but it also realized that the Cape's past was "an important one too."[71] In this early "historic preservation" campaign the Cape Cod Improvement Association worked to find a resolution to the conflict between development and the Cape's historic past, hoping for "public improvements . . . with the minimum destruction of features of natural beauty." The association shared Perry's belief that Cape Codders should embrace a future based on tourism and publicize its "unsurpassed bathing facilities, its beaches . . . of clean white sand, washed by waters clear, clean and warmed by the gulf stream . . . its boating and fishing facilities and . . . [a] locality about the most healthful in the world."[72] But its members were also concerned that Perry's focus on making money selling Cape properties could also lead to the Cape's destruction.

The association reflected the concern that the private profit of the developer should not be allowed to undermine the public interests of the towns and their citizens. Once again, after a hundred years of little debate or conflict, clashes between private interests and the public good of the town became a concern on the Cape. Whereas seventeenth- and eighteenth-century Cape Codders struggled with the issue of protecting the commons from overharvesting of lumber, overgrazing by privately owned animals, overfishing, and destruction of fish migration, their twentieth-century counterparts struggled to protect town access to beaches and waterways, scenic views, and open space. "All public issues should be settled by each person on the ground, not as to what makes for his own private interests . . . but as to what makes for the permanent good of the town."[73]

James Nickerson of Harwich led the Cape Cod Improvement Association and acted as its first president. As an important local business leader and with a family history that stretched back to the early days of settlement, his vision of controlled development and transition from the past to the future carried weight. Ezra Perry, however, recognized the conundrum: the larger economic pressure to "make money for the land is increasing in value very rapidly each year," he wrote, focused attention on "developing [the towns] as summer resorts" without thinking about the "permanent good of the town."[74] With the Cape still reeling from declining fishing and manufacturing and a shrinking population, protection of the "past" seemed irrelevant. Shortly after Nickerson's death in 1906, the Cape Cod Improvement Association collapsed.

THE AGE OF THE AUTOMOBILE

With the coming of the new century, a new means of transportation, the private automobile, began to affect the lives of Americans. Although plagued by bad roads, automobile travel opened up the Cape and was as revolutionary as the railroad. As the coming of the railroad ended the Cape's amphibious culture, the private car ended the era of steam.

At first the automobile was the plaything of the rich. Although it freed its owner from the difficulties of moving trunks and luggage from home to train and train to hotel, it was by no means without its problems. In 1904 it took an entire day to travel from Boston to Yarmouth by car. Cape roads were mostly sand and only slightly improved over those about which Henry David Thoreau complained.[75] It was not until the 1920s that the car radically transformed vacation habits and Cape Cod.

In the 1890s—in response to the growing enthusiasm for bicycle riding, an interest in facilitating the movement of Massachusetts-produced goods, and to "serve the interests of pleasure travel"—Massachusetts set up a highway department and began a program of highway construction and improvement by covering roads with asphalt.[76] The state, anticipating the growing tourist economy, noted as early as 1898 that "a considerable part of the present and much of the prospective value of real estate in the Commonwealth . . . is due to the incoming

Trains, Cars, Cottages, and Restaurants 129

of people from the central and western parts of the country who seek summer residences and family homes in the very attractive rural districts of the State. People who thus resort to the Commonwealth for recreation desire opportunities for driving such as would be afforded by a well-organized system of State ways, which would be laid out and beautified with the natural and historic interests of the country."[77] In view of its belief in the importance of highway construction to entice tourists to the state, the Massachusetts Highway Commission spent $42,992.11 in its first four years in Barnstable County.[78] That amount was the third-largest the commission allocated to a state county; the same tourist-motivated considerations encouraged local communities to build and improve their roads by taking advantage of the infusion of tax revenue from the large estates rising up along the coast.[79] In 1913 the Cape began the process of tarring its roads. Between 1890 and 1921 the Massachusetts state government spent $826,310 on Cape highways and contributed another $1,353,355 to towns for road construction.[80] By the 1920s the Cape had 136 miles of paved state highway and 1,100 total paved miles. This was considerably more than in similarly populated areas.

Even with the new state and local roads, E. C. Janes remembered the family trips from central Massachusetts to Wellfleet beginning in 1913 as ordeals, with the family leaving for the Cape in the touring car at five-thirty in the morning and not arriving in Wellfleet until the evening. With the car loaded down with a steamer trunk and several suitcases, two spare tires, and an extra five-gallon can of petroleum, the family had to send on ahead "two large wardrobe trunks filled with sheets, pillowcases, towels, bathing suits, dress suits and other stuff" by railway express.[81]

With the coming of the automobile, the large hotels and vacation resorts, such as the Highlands House in Truro, gave up offering meals to their guests and built kitchens in the cottages. Guests then used their car to shop at the roadside stands, fish markets, and country stores for the makings of their own meals.[82]

The increased use of the automobile as a means to arrive at the Cape freed vacationers from the anchor of the large hotels and their livery service. Visitors could ship their larger trunks to the town depot, load the car, and then head toward any Cape destination their automobile could negotiate. Agnes Edwards remembered trips to the Cape before automobiles as being major projects. The wealthy spent days packing up their luxuries in the city before heading down to their "charming summer places," with their "air of feudal estates."[83] But within a few years, "the automobile . . . brought the summer cottage within a few easy hours of Boston. Today [1918] on Saturday afternoons, the well oiled highway is alive with cars and on the great drawbridge over the canal [opened in 1914] . . . two traffic policemen are kept busy from dawn to dusk."[84]

Once at the cottage or modest guesthouse, visitors unloaded their cars, made a quick trip to the town depot, and settled in for vacation independent of the services of the hotels. Their cars also opened up more of the Cape. Other towns could now be easily visited. Roadside fruit and vegetable stands could be sampled, and, if the weather did not favor a trip to the beach, one could take a trip to the interior of the Cape to look for local antiques, out-of-the-way picnic sites, or secluded ponds.[85]

The automobile allowed visitors to appreciate the Cape's variety. The automobile also affected the transformation of the Cape economy far more than did the railroad and the vacation hotels. For those coming to the Cape the car meant mobility; for those on the Cape it meant new opportunity.[86] Homeowners whose attics and barns had been accumulating stuff for generations now found a market for what previously had been flotsam from the sea or old possessions not yet thrown away.[87] Hungry tourists, traveling about the Cape in cars, offered a new market for enterprising locals who diverted oysters, clams, fish, and lobsters meant for off-Cape markets to roadside cafes and restaurants.

THE FASHIONABLE SET

Although the automobile opened up the Cape to the growing American middle class, it did not halt the continued development of the large estates. Falmouth, Hyannis Port, Cotuit, Osterville, and Chatham continued to see more and more of their prime beachfront or water-view property converted into estates for the very wealthy. Already by 1890 more than a fourth of Falmouth's town taxes were paid by four wealthy families who financed the building of "elegant residences."[88] The Chapoquoit area of Falmouth became one of the most fashionable resorts on the Atlantic seaboard by the early twentieth century. And not all the estates were summer homes. Retired businessmen and successful investors and lawyers from Boston, New York, Philadelphia, and Chicago increasingly transformed their residences into permanent homes. By 1914 Falmouth was the sixth-wealthiest town in Massachusetts.[89] Because Falmouth developed early as a resort community, centered on large estates held by wealthy families, a large part of the town remained undeveloped in the form

Trains, Cars, Cottages, and Restaurants 131

Cape Codders increasingly catered to the new tourists. This old shop on the edge of the shore replaced marine gear and fishing tackle with items that appealed to the new Cape visitors. Photo courtesy of the Library of Congress.

of land that stretched out behind the large homes hugging the coastline. The undeveloped land of these large estates helped preserve the exclusive nature of the town while development in other towns began filling in empty land with cottages and rental units.[90] Into the 1930s Falmouth was "a lively fashionable [town] of summer life centering around golf courses, yacht racing, [and] polo matches."[91] Although summer visitors drove the Falmouth economy, the development and expansion of the science centers in Woods Hole also contributed to the town's cultural allure while adding employment opportunities. In 1871 the U.S. Commission of Fish and Fisheries set up a summer station in Woods Hole that expanded into an aquarium and permanent fish station. The presence of this scientific center encouraged scientists to establish the Marine Biological Laboratory there in 1888 in an old whale-oil candle factory. These institutions drew in dozens of scientists and gave the community an intellectual cast, which helped sustain other activities within the town such as concerts and lectures. At the same they provided employment for local residents who maintained the institutes.[92]

By the late 1920s the Cape offered a patchwork of vacation opportunities. The exclusive areas on Nantucket Sound from Falmouth to Hyannis Port offered yacht clubs, tennis courts, and private beaches to those who built, bought, or rented homes. The villages of Hyannis Port, Craigsville, Centerville, Osterville, Oyster Harbors, Wianno, and Cotuit were "fashionable summer resorts attracting wealthy visitors. Activities centered around . . . yacht clubs, fine bathing beaches, [and] big hotels." The warm waters of Nantucket Sound were "alive with sail" where "big yachts swing at anchor." Towns along the Sound were places where "people come and go by plane or high-powered car [and] city orchestras rather than radios play for dancing at night." Unlike the towns further down the Cape that were also filling up with small cottages, these towns of "big luxurious Victorian summer homes, set in carefully landscaped gardens, and their wide-veranda hotels" became places of play for the very exclusive set.[93] When developers attempted to bring in popular "amusement" to Falmouth, town leaders denied permission, fearing that the amusements would attract less affluent vacationers such as those who flocked to Nantasket Beach, south of Boston.[94]

Golf became a popular sport on this side of the Atlantic in the 1890s as wealthy Americans began imitating the life style of the British upper class. Homes were built to imitate the estates of the landed English families. Even the architecture of colleges where wealthy Americans sent their children increasingly resembled that of Oxford and Cambridge. Foxhunting clubs sprang up in suburban Philadelphia, New York, and Boston, and the game of golf became the preferred sport of wealthy Americans. At the turn of the century, the most famous golf courses were along the coast of England, Wales, and Scotland. These became the models for American courses.

To developers who eyed the Cape for its potential, the eroded and sand-blown farmland that had so distressed late-nineteenth-century investigators looked like the landscape of a Welsh, Scottish, or English golf course. By the beginning of the twentieth century there was an eighteen-hole course in Woods Hole and another one at the Wianno Club linked to the Wianno Yacht Club. Sesuit Club boasted a nine-hole course for its guests. Sagamore had a nine-hole course, whereas the more exclusive Hyannis Port featured an eighteen-hole course. There were nine-hole courses in Cummaquid, Pocasset, and Harwich. Yarmouth claimed that its eighteen-hole Bass River course was one of the oldest in the country. The Chatham Bars Inn operated a nine-hole course while the Chatham Country Club boasted of its world-renowned eighteen-hole champion course, "Eastward Ho," designed by Britain's most famous course designer.[95]

Those with means not only built communities where they could recreate with those of similar interests and social status, but they also sent their children

Trains, Cars, Cottages, and Restaurants 133

off to camps where the youth of the nation's elite could meet and play with peers from around the country. Cape Cod's growing cachet as a playground for well-to-do adults, combined with the available and inexpensive land prices, encouraged the establishment of summer camps for children. In 1900 Portanimicut Camp for boys opened. The camp, catering to young, upper-class East Coast male children, offered its "Bell Boys" swimming and sailing activities away from the heat and crowds of the city. The success of Portanimicut Camp led to a rash of summer camps opening up across the Cape over the next thirty years, offering swimming, sailing, archery, fishing, and boating in both male and female camps. Pleasant Bay alone had five sailing camps. In 1905 Mary Hammatt opened the Quanset Sailing Camp, which continued for seventy years. Alice Hamilton, who taught at the prestigious southern women's college Randolph Macon, set up a nautical camp for girls in 1914 in Wellfleet, while the one-hundred-acre Camp Cowasset, an exclusive girls' camp in Falmouth founded in 1915, drew in ninety girls aged six to eighteen. In 1925 Alice Murdoch started Camp Cheri for young girls, adding French to the usual fare of swimming, rowing, sailing, leather working, and basketry. Four years later Camp Avalon opened up for young girls not interested in speaking French. Summer camps were thickest on the ground around Pleasant Bay, but they could be found in most Cape towns. Although a 1926 Cape newspaper argued that the summer camps were "not economically sound," being "isolated [and] located on extensive tracts of otherwise vacant ground," most lasted into the late twentieth century.[96] These camps were not the summer places for the lower classes. In 1923 Camp Quanset, for example, charged the parents of campers $350 for the summer, or $180 to $185 for half the summer season, at a time when a well-paid skilled worker would have been lucky to earn $200 in salary for two months of work.[97]

Although more and more residents were coming to the Cape by car, the railroad continued to be a means of getting to the Cape, especially for weekends. The *Cape Codder* left New York on Monday through Saturday mornings and arrived in Hyannis in the afternoon. It left Hyannis each morning for New York City. In 1944 the New York–New Haven–Hartford Railroad instituted the *Neptune,* a deluxe weekender that left New York on Friday afternoon and arrived in Hyannis at nine-thirty at night with bus connections to the rest of the Cape. The train left Hyannis at midnight on Sunday and arrived in New York City on Monday morning in time for work. Families were packed off to the Cape for a month or two while the breadwinner continued to work in the city during the week and traveled to the Cape for the weekend. Boston's regular train service also allowed the family to take the car to the Cape while one member could stay behind working in the city and travel down on weekends.[98]

The wealthy continued to build summer estates in the more prestigious towns and communities, but the large luxurious resort hotels began to suffer by the late 1920s. The wealthy increasingly abandoned these hotels for their own estates, and for the middle class, the hotels had always been a financial stretch. The explosion of small cottages and tourist homes available to those with access to an automobile and the increased practice of weekend visiting offered vacationers a cheaper alternative to the large hotels. The Chequessett Inn in Wellfleet, which offered comfort and a variety of social activities, prospered through the 1910s and 1920s. The Inn offered accommodation in cottages as well as in the main building. While the cottages remained popular, the inn began to struggle by the end of the 1920s. In fashionable Falmouth, the Cape Codder, probably the Cape's largest hotel with 125 rooms, charged five dollars a night on the American plan. Increasingly it also had trouble filling its rooms. Built in the days when visitors arrived by train with packed trunks and were picked up at the station by a hotel coach for an extended stay of a week or more, the hotel found it difficult to convert to visitors arriving by car for short stays of a weekend or even a day.[99] The coming of the Depression hurt most Cape tourism, but it hurt the large hotels, such as the Cape Codder, the Chequessett Inn, and the Nobscussett House the most. When ice floes damaged the pier beneath the Chequessett Inn, it was not repaired, and the inn was torn down. Tourists kept coming to the Cape even during the lean years of the Depression, but they came on tight budgets. Camping, cheap cottages, or tourist homes were what they could afford.[100]

OPPORTUNITIES IN RENTING, SERVICE, AND CONSTRUCTION

Those who remained on the Cape found new opportunity in empty rooms. E. C. Janes and his family, who came down from central Massachusetts, stayed in the home of the Hopkins family. Typical of many Cape natives, Mr. Hopkins went to sea at twelve years old as a cabin boy on the Wellfleet schooner *Unis P. Newcomb*. He was promoted to first mate by his mid-twenties, but the decline of the fishing industry forced him to leave the Cape in search of employment, which he found as a salesman in Springfield. With the coming

Tourists from Boston come into Provincetown by boat. Although traditional Cape activities continued, by the twentieth century tourism was replacing fishing as the economic lifeblood of the Cape. Photo courtesy of the Library of Congress.

of the automobile to the Cape, Hopkins returned to Wellfleet and converted the family home into a two-family structure, renting out the downstairs to summer visitors while his family of four lived upstairs. In addition to renting out his home, Hopkins did odd jobs for summer guests, including arranging for horse-drawn buggies to take visitors to non-automobile-friendly out-of-the-way places, such as Gull Pond for fishing.[101] Hopkins also connected guests with local fishermen who took summer visitors out for deepwater sport fishing in old catboats. By the early years of the twentieth century these deepwater fishing trips became extremely popular with male vacationers. Bluefin tuna, the bane of commercial fishermen in the nineteenth century because they tore through pound nets and drove away more marketable fish, were transformed into prize sport fish in the twentieth century. Fishermen who had cursed tuna a half century earlier now bragged of their ability to locate the exciting fish.[102] Swordfish, which were left on the pier in the late nineteenth century, became sport fish in the twentieth and were sold to hotels and restaurants for ten cents a pound by 1910.[103]

Tourists brought more opportunities to Cape residents who were developers, fishermen, farmers, or families renting out rooms or cottages. It also brought employment in service and building trades. The large estates had huge gardens and lawns that needed tending, and the homes needed constant repairs after winter weather took its toll. Although some farmers were able to expand their operations into successful truck farms, others gave up trying to coax sufficient produce from the sandy soil to make a living. Even successful farms would often need extra income to stabilize the family's finances. Caring for, maintaining, and repairing summer homes and estates for nonresident owners became an important part of household income for Cape families.[104] George Ellsworth, a son of a Cape lighthouse keeper, took one of the few jobs available for young teens in Chatham. He worked for the Monomoy Life Saving Station. Ellsworth put in seven hard years rowing out in perilous weather to save victims of shipwrecks. After putting in his time at the station, Ellsworth, in 1920, "turned to another local opportunity, working for the rich." In that capacity he acted as a year-round caretaker for a Chatham estate and "took out fishing parties and guided duck hunters."[105]

The wealthiest Cape visitors brought their servants with them when they arrived in the summer, but even these families needed their estates looked after when they were away.[106] Boats and yachts needed dry-docking. Leaves needed to be raked and gardens put up for the winter. Pipes needed draining, mice were to be trapped, and shutters had to be closed tight.[107] Once on the Cape, even those who brought their own servants often needed extra help in maintaining their households, oftentimes greatly expanded by guests. Cape residents not only supplied food for restaurants, hotels, and guesthouses, but many of them, mostly women, also worked preparing that food. The demand for summer help far outstripped the local labor market, and supplemental summer workers, particularly students and recent high school and college graduates, added to the influx of summer employees. Falmouth's working population doubled over the summer months and of those who worked summer jobs, two-thirds were not from Falmouth. College students made up the majority of those employed in stores and restaurants, and even the local dairy employed college students to deliver milk during the busy summer season.[108]

The emergence of cottages focused more on middle-class tourists and shifted the employment pattern for those who catered to the summer trade. Work for gardeners and year-round permanent caretakers, many of whom lived on the estates, declined in the period after 1920. The caretakers' positions were replaced by summer cleaners, part-time maids, cooks, and staff in motels, hotels, restaurants, and summer cottages.[109]

SUMMER COTTAGES AND MOTELS

With the collapse of traditional Cape occupations, shorefront property prices remained low into the twentieth century. Although the most desirable land along the more prestigious Buzzards Bay and south shore was already developed by 1920, there were still thousands of acres of undeveloped land, either abandoned fish-drying and salt-making areas or abandoned farmland and woodlands. Driven by new cheap cottage construction and initially cheap land, developers moved onto the Cape in a rush in the 1920s. Land speculation, especially on the south shore from Falmouth to Orleans, caught the imaginations and pocketbooks of both Cape and nonresident investors. Some land was bought, divided, and sold off simply as speculation. Along with land speculation came development projects.

Areas such as the cottage villages that were springing up, along the shores of Nantucket Sound from West Yarmouth through Dennis and West Harwich and the bay shores of Eastham, Wellfleet, and Truro, catered to the emerging middle and lower-middle class. These areas offered renters and small cottage owners the opportunity to walk, wade, swim, and fish, set up badminton or croquet matches, rent a rowboat, or just sit on the beach. Although the wealthy purchased the land of abandoned boat works, saltworks, and fishing wharves along Buzzards Bay and Nantucket Sound, developers and owners of modest parcels built small cottages and rental units. The shoreline to the east of Hyannis Port was divided up into smaller lots and sold to those with limited budgets and modest aspirations for their summer cottages.[110]

The Orleans Associates, made up of local businessmen, purchased the Mayo farm at Pochet in East Orleans, laid out lots, and built model cottages. Unfortunately for the Orleans Associates, their plans were too ambitious for even the prosperous 1920s, and they went bankrupt in 1928. Joshua Nickerson, hoping to do better, bought up the development at a foreclosure sale. Offering lots for $675 apiece, Nickerson could not even sell enough lots to meet his mortgage and taxes until the late 1940s. It was not until the late 1950s that the lots sold for the price the Orleans Associates sought in the 1920s.[111]

In 1935 the federal Public Works Administration (PWA) built two bridges across the 1914 canal, and motels and motor courts soon sprang up across the Cape, especially along Route 28, a south shore road that had been paved before World War I. By the 1940s many residents, originally drawn to the Cape by its summer beaches and beautiful vistas, were winterizing summer cottages or buying all-season homes for retirement. Developers, realizing this new market, searched out cheap open land back from the shore, behind the expensive coast property, and began building clusters of winterized houses catering to retirees.[112] Although the increase in tourism led to a resurgence of Cape farming, the demands of land use pushed against farming. By 1935 the value of land for tourism (subject to yearly assessment) on the Cape had risen to over seventy-five dollars per acre.[113] Initially, in the early part of the century, the land purchased and used by vacationers was along the coast and therefore not suitable for asparagus or strawberry farming. But by the 1920s the Cape's resident population grew with retirees or people who wanted to take advantage of employment opportunities of the expanding vacation economy. Most of these residents wanted homes but could not afford or did not desire to be right on the coast. They looked inland for housing and in doing so competed with use of the land for agriculture. By 1940 only 14 percent of Cape land was farmed as opposed to 18 percent in 1920. By 1950 farmed land had fallen to 8 percent.[114] Agriculture brought in only 3.5 million dollars to the Cape in 1950. Land that catered to tourists and was suitably sited for building shopping centers, homes, and restaurants brought far more money to the Cape than that which produced broccoli, lettuce, and asparagus.[115]

Although fishing, farming, and boatbuilding had continued on the Cape into the twentieth century, and even experienced a certain resurgence after World War I, the future clearly lay with tourism. By 1935 over twenty-five million dollars was being spent on the Cape by summer people.[116] Summer visitors accounted for 54 percent of Falmouth's economy by 1940.[117] Its permanent population of seven thousand in 1940 depended upon the summer trade for 75 percent of its annual income.[118] By the 1940s nonresident summer people were paying the majority of real estate tax, contributing over $2.5 million to Cape towns.[119] In Chatham, nonresidents paid a combined $75,917 compared to $59,650 from year-round residents, and as early as 1925 summer residents—some recognized by name or reputation, such as Louis Brandeis, an associate justice on the U.S. Supreme Court—owned one-third of the houses in Chatham.[120] In 1950 tourism accounted for over seventy million dollars to the Cape's economy, while fishing, agriculture, and manufacturing combined added only eight million dollars.[121]

In 1920, although the famous Nobscussett House along the north shore still brought significant numbers of visitors to the town, the residents of Dennis still depended upon fishing and agriculture, primarily in cranberries, eggs, vegetables, and a very limited amount of dairy products. The town had no fire or police departments, no electrical service, and no municipal water service. By 1950, the dairy farms had closed and the town's vegetable and egg production had fallen to insignificance. Competition with off-Cape cranberry producers in neighboring Plymouth County and those in New Jersey and Wisconsin had marginalized even the crop that once had been a mainstay of the Cape's agricultural economy. Bumper crops from 1947 through 1951 drove down prices from twelve dollars a barrel to five dollars. Cape cranberry producers were also hurt by their small, scattered bogs, which rendered mechanical harvesting impractical. Dennis farmers, sharing these problems with other Cape cranberry producers, increasingly gave up cranberry raising altogether, filled in their bogs, and offered the land for sale for development.[122] The Nobscussett House closed its doors in 1930 to be replaced during the next twenty years by dozens of small cottages, motels, and guesthouses that sprang up along Route 28. By 1950 the town boasted of a summer theater, a cinema, and, reflecting the new automobile culture, a drive-in movie theater and a trailer park.[123]

The cottages and motels that began to crowd in along the coast from

Hyannis to West Harwich sprang up with amazing speed as Americans came to enjoy more discretionary income and leisure time. The successful unionization of a large part of the workforce and the passage of the Fair Labor Standards Act fostered the forty-hour workweek and paid vacations. More and more Americans loaded their cars and headed to the beach, and with the growing crowds of visitors, Cape Codders continued to sell their land in small lots to these newcomers and to developers who bought up lots and subdivided them into small cottage lots. Locals and newcomers looking for employment offered their labor and skills to build the cottages to fill these lots. Some new lot owners ordered their cottages from Hodgson Homes, which offered to deliver prefabricated cottages that could be simply bolted together on the lot.[124] Once the house kit was delivered to the lot site, the owner would either hire local labor to set down cinder block posts and bolt up the cottage or bring a group of friends to help do it. Although the postwar prosperity hastened this process, it had begun as early as the 1920s.

In 1922 a state report noted that "numerous summer cottages line the miles of sandy shore, and the land for some distance back from the shore is rapidly being built upon by those desirous of securing a permanent summer residence."[125] Although it was the south shore between Falmouth and Harwich that saw the most development of small cottages, other towns also saw small summer cottages spring up where previously there had been decaying wharves, abandoned saltworks, and fish flakes. By 1920 much of the shore property along the bay in Truro was owned by nonresidents, with the value of land increasing "greatly in recent years and already summer cottages are being located on inland hilltops."[126]

In 1926 Samuel Hannah, originally from Ohio, moved to Sandwich and organized a development company, Homestead Trust. Hannah bought thirty-six acres from the summer estate of Edwin Blake, who himself had bought the land in 1900 from an old seafaring family that had fallen on hard times. To this property, Hannah added some unclaimed land he found in searching through the County records. He then laid out three hundred home sites that were 125 feet deep with fifty-foot frontages. These were lots for small cottages, not huge estates. The Depression slowed down Hannah's project, with most of the lots not selling until the 1950s, but the pattern was clear: Cape Cod development in the age of the automobile would result in more small cottages than large estates.[127]

The small cottage construction on small lots with individual septic systems was the development trend from the 1920s, particularly along the south shore. Although this trend offered opportunity for more construction and more jobs, not all Cape Codders saw it as positive. Joshua Nickerson II, whose own father had bought the bankrupt Orleans Associates development project and offered small lots at $675 a piece in Orleans, was one. Chatham had been fortunate, he commented, that developers like Charles Hardy had developed land in large lots and pushed up the land values so that "tiny cottage lots as happened in old camp meeting sites such as . . . Harwichport and Eastham" were avoided by Chatham.[128] By the 1940s, even in Falmouth, some of the land surrounding the large estates was sold off into smaller cottage units or for commercial recreational use.[129] Assessed land value climbed in Falmouth through the early years of the twentieth century from $928,155 in 1900 to $2,305,545 in 1920, and it reached $5,919,054 by 1929. The dramatic increase in land prices encouraged landowners to sell. Most of those who initially gained from the increased land prices were external speculators, but by the 1920s everybody seemed to get involved. Any empty spot of land was bought up for development and speculation. As more people flooded onto the Cape, small lots were carved up and rental cottages built.[130] Not only the automobile but also new cheap cottage construction and small lots opening up for those cottages encouraged the influx of hundreds of new visitors and summer residents to Cape Cod. So did the emergence of new forms of recreation and sport that were embraced by the middle and upper middle classes.

Once at the cottage, the new visitors faced the problem of what to do. Nineteenth-century upper-class Americans had perfected the art of leisure. At their summer estates in Bourne, Falmouth, or Chatham they took strolls, contemplated or painted watercolors of scenery, sailed, golfed, or socialized with others of their class for singing, reciting, or playing cards. The American middle class in the early years of the twentieth century was just coming to grips with the idea that leisure, which previously had been scorned as time wasted, was a positive rather than a negative quality. These Americans did not have the resources to buy and maintain yachts or belong to country clubs, nor did they have the patience for strolling, painting watercolors, or reciting poetry, and playing cards was still a suspect activity. All that changed in the aftermath of World War I. By the 1920s the American middle class embraced leisure.[131]

Public swimming or at least vigorous wading, fishing, building castles in the sand, rowing simple skiffs, and evening board games and cards became acceptable activities.[132] These were activities that middle-class visitors to the Cape could embrace. Small cottages seemed less cramped if more time could be spent being active outside. The two or three weeks on the Cape were now filled with activity for the whole family.[133] The new Cape visitor came by car and could move across the Cape looking for things to do and things on which to spend money. Some of this money went to renting fishing boats, rowboats, or sailing skiffs, or buying fish and vegetables for dinner. Some of it went to buying souvenirs or antiques from roadside stands and back road cottages with merchandise spread along the front yard and a "for sale" sign attached.

The new cottages also benefited from the arrival of electricity in the 1910s. Renting a cottage at the turn of the century had involved hand-pumping water, bringing or buying kerosene for lighting, and working a wood stove to cook.[134] For the wealthy, coming for an extended summer stay, these problems were overcome by a staff of servants. For the renter or owner of a small cottage, bringing in kerosene, hand-pumping water, and firing up a wood stove in the heat of the summer were unwelcome chores.[135]

Cape Codders not only opened their homes to guests. Those who held land near drivable roads built simple one- to three-room cottages that they let out for the week or month. Others, in imitation of the older revivalist campgrounds, built outhouses and cleared out brush to rent to campers, who shipped camping equipment by railroad express, loaded down their cars with what they did not ship ahead, and headed to the Cape for an extended camping experience.[136] Christopher Winsor remembers his family driving two days from New Jersey to camp in Wellfleet in the interwar years. Winsor's family joined others who camped because it afforded them more freedom and was considerably cheaper than renting a cottage or staying at a boarding house.[137]

The replacement of the American plan resort hotels by small cottages and motels gave rise to a new industry on the Cape, the resort restaurant. The large hotels and guesthouses supplied all the meals that their patrons desired for the single cost of staying in the hotel, and the large estates had their own in-house cooks, but cottagers and motel visitors now looked to local restaurants for some of their fare.

PROTECTING THE CAPE IMAGE

Realizing that the Cape was rapidly changing and wanting to both profit from that change and control it, Cape business leaders came together in 1921 to form the Cape Cod Chamber of Commerce. The organization set the promotion of Cape Cod as its goal, which initially meant accentuating traditional economic activities of farming and fishing as well as the new tourist economy.[138] Following the push by the Cape Cod Farm Bureau, the chamber also championed the building of a freight terminal on Buzzards Bay so that fishermen and farmers could get their produce to distant markets. The fate of the freight terminal reflected the larger story of Cape Cod. By the time it was completed in 1935 there was not enough produce to ship out, and the pier was turned over to the Massachusetts Merchant Marine Academy.[139]

Although the Chamber of Commerce supported a freight terminal to facilitate moving Cape seafood and agricultural produce to off-Cape markets, by the late 1920s its focus had shifted to the vacation economy. In 1928 hotel and restaurant leaders persuaded the chamber to launch a $35,000 publicity campaign to sell the Cape as America's vacationland.[140]

Although the Chamber of Commerce supported the emerging Cape tourist economy, its members found a cause for concern in the growing number of the small cottages. In 1926 the chamber invited Garret A. Banneker, a nationally renowned artist and summer resident of Truro, to speak at its Hyannis meeting. Banneker claimed that people came to the Cape for its natural beauty but also for its "quaint atmosphere and the spirit of our early settlers. There is charm about Cape Cod which is to be found nowhere else save in New England and this charm lies in the quaintness of the little homes and churches built by hands and simple spirit of our forefathers many years ago." Banneker warned the chamber that if their only interest was "in selling real estate, in turning over acreage to your best immediate advantage regardless of who is going to buy these acres and build upon them, if you are interested only in getting swarms of people to come here—which will in turn bring more business . . . your project will fail for Cape Cod will cease to attract."[141] Banneker argued that if the chamber should "foster the preservation of traditional atmosphere and spirit of Cape Cod and promote new development and building in keeping with this historic quaint and charming spirit, you will find yourselves building on a foundation which will constantly tend to enhance and increase not only the desire of appreciative people to come here but you will also increase the value of your own property as well."[142]

Banneker and the chamber, like James Nickerson's Cape Cod Improvement Association a quarter-century earlier, were concerned about the direction and speed of the emergent tourist economy. But both Banneker and the chamber focused their concern on versions of class privilege. The rapid expansion of the Cape's tourist economy initially did not threaten the older regime, but rather benefited from its decline. The decline of the regime of production and resource extraction opened land for other uses, and those other uses also provided new life to elements of the older industry. Economic tensions between the older and newer regimes were building as land prices increased, but land prices were not the issue on which Banneker and the chamber focused. They romanticized the older regime and worried about the devaluing of the product with too many people coming to use it. They saw the Cape's traditional heritage as a selling point and worried that too much development would undermine the image of the Cape they hoped to project to potential visitors.

In 1927 the Chamber of Commerce hired regional planner Benton MacKaye to advise on what the Cape should do to protect and expand its tourist industry. MacKaye, like Banneker before him, stressed controlled and planned growth.[143] MacKaye pointed to the importance of locating activities in appropriate areas, clustering commercial districts, and maintaining green space. But the message the chamber took away was one of protecting the Cape from the wrong kind of development. Ironically as the Cape was becoming a vacation destination for the middle and upper-middle classes, and as its shorefronts began to fill in with small cottages and motels, the chamber launched its Cape Cod Advancement Plan in 1934 to publicize the Cape as a vacation area for the nation's wealthiest elites. It published two books focusing on the Cape's history of fishing and artisan production. It stressed that the Cape was populated by old-stock Protestant Yankee farmers and fishermen who had old-fashioned charm and character. It also pointed to the Cape's beauty and the existence of both its exclusive wealthy communities and its transportation links to the eastern U.S. cities.[144]

Whether they were wealthy or not, tourists came to the Cape for its beauty but also to have fun. And by the 1920s having fun involved spending money. Some of that spending went to purchase items shipped in from the very places the vacationers were escaping. By the 1920s Hyannis had emerged as the commercial center of Cape Cod. It not only sported the largest number of Cape restaurants and the region's only hospital, opened in 1920, but it increasingly had branches of major department stores, such as Jordan Marsh and Filene's. Other towns also saw a growth in retail stores catering to those with off-Cape tastes. In 1940 Falmouth had a branch of Filene's and a Howard Johnson restaurant that were open for the summer months.[145]

The new tourist economy depended on visitors with enough discretionary income to take a vacation. Much of that discretionary income evaporated between 1929 and 1933, and unemployment on the Cape surged to 38 percent. The Cape's population, which fell 26 percent from 1860 to 1920, had finally grown in the 1920s, but fell again over the 1930s. Cape leaders pushed for public projects to increase access to the Cape. With the backing of the Chamber of Commerce, the New Deal Public Works Administration built two bridges across the canal, dramatically increasing the ability of visitors to get on and off the Cape. By 1937 a visitor to the Cape could travel from Boston to Orleans by car in as little as two-and-a-half hours.[146] A Civilian Conservation Corps (CCC) project on the Cape helped reforest the new 8,500-acre Shawme-Crowell State Forest on the edge of Bourne and Sandwich and reworked the land the Nickerson family gave to the commonwealth into a state park, the first in Massachusetts. The new Nickerson State Park offered camping facilities that gave those on a very limited budget, but with access to an automobile and tent, the option of vacationing on Cape Cod.[147] Although the nation continued to struggle economically, the worst of the Depression was over by 1935, and visitors once again flocked to the sandy shores.[148]

By the end of the 1920s, Cape boosters such as the Chamber of Commerce not only believed that tourism represented Cape Cod's economic future but were also determined to take steps to make the Cape even more enticing to potential visitors. That involved, among other things, addressing the problems posed by mosquitoes, greenhead flies, and other insect pests that had plagued Lorenzo Dow Baker and led him to push the town of Wellfleet to dike its river. The dikes were designed to reduce the breeding area of the salt marsh mosquito, *Aedes solicitans,* making coastal property more appealing to tourists and land buyers. Unfortunately for the town of Wellfleet, the dikes, including the one at the mouth of the Herring River, did not eliminate the problem of mosquitoes. Wellfleet was not alone in struggling with the problem of this biting insect. Most Cape towns suffered from these pests. The grazing animals' pastureland returned to brush and scrub forest, which provided good habitat for mosquitoes. Abandoned cranberry bogs and stagnant water behind the railroad beds and roads cut off the natural flush of tidal water into wetlands and provided breeding grounds for insects. Continued complaints that mosquitoes were driving away visitors, and that they represented a public health menace, led individual towns to enact mosquito abatement programs that involved oiling wetlands and ditching and draining marshes.[149] Pressure from developers and new landowners "holding . . . the best shore property since the decline of fisheries" led Wellfleet to appropriate fifty thousand dollars on mosquito abatement over the ten years after building the dike at the mouth of the Herring River.[150] In 1928 the Chamber of Commerce launched a Mosquito Control Project, and in 1930 Massachusetts enacted a law providing for the creation of county mosquito control projects. The keystone to these projects was wetland drainage, facilitated by digging channels through wetlands to drain water. Initially funded by the state, with the creation of the WPA, these projects became large-scale public works endeavors, eventually digging over one thousand miles on Cape Cod.[151]

Mosquitoes may have been seen or felt as one of the Cape's biggest problems, but they were not the only problem plaguing the region. Mosquitoes were natural to the Cape, but the humans who were annoyed by the mosquitoes caused many of the Cape's problems. Although low population protected the Cape water system from the types of pollution plaguing urban water supplies, the more densely settled and built-up towns such as Falmouth found that a lack of clean water was a problem as early as the turn of the twentieth century. Private wells were increasingly considered inadequate to supply the water needed for fire protection and household consumption. By 1902 Falmouth center was served by a municipal water system drawing its water from Long Pond. Although most of the townspeople found their water to be excellent, concern was raised in the 1920s about sewage. Without a town sewage system, residents dumped their wastes into cesspools. Increased population density and the ensuing increase in number of these cesspools led to arguments for a municipal sewer system. By 1927 Eel Pond was compromised by seepage of sewage into the pond, but the town voted down plans for a sewer system. Those who lived outside the town center believed their cesspools perfectly adequate. Continued episodes of waste seeping into ponds led to renewed calls for a treatment plant, but the cost of $110,000–$125,000 in the midst of the Depression led again to the town voting against a system in 1931.[152] Sewage was only one of Falmouth's waste problems. By 1940 the town dump was approaching full capacity, and the town was casting about for an alternative site for its garbage.

Falmouth, with its higher population density, faced the problems of waste early in the century. In 1940 garbage and sewage problems seemed to most Cape Codders as local problems and only of concern to the few towns that had significant urban densities. That was not the condition for most of the Cape on the eve of American involvement in World War II.

POSTWAR PROSPERITY

Although the Cape recovered from its long economic decline in the 1920s, the Great Depression of the 1930s threw the Cape once again into troubled economic times. But by the late 1930s, towns on the Cape began recovering their populations, as work to improve access to the Cape encouraged its recovery.[153] In 1935 New Deal programs built two multilane bridges across the Canal. The same year the government bought up land in north Falmouth, available because of the failure of the huge Coonamessett Ranch, and constructed a National Guard Training Camp.[154] In 1940, with the growing possibility of war, the federal government spent twenty million dollars to refit the area into the Camp Edwards military base. In addition to hundreds of housing and storage buildings, the government built a hospital, a sewage treatment plant, and railroad lines and highways on to and through the base.[155] Construction employment on the Cape jumped and unemployed workers from around New England traveled there to find work.

The end of World War II not only brought home returning GIs and reunited families, but it also launched the nation on its longest and most profound era of growth and prosperity. That prosperity saw the massive explosion of the housing industry, and the creation of modern suburbs linked to private transportation. On the Cape it brought ever more tourists, vacationers, and retirees. It also brought more homes, cottages, and motels, and mile upon mile of highway. George Redding, an old retired professor living on the Cape, commented at the time that it would not be long before the Cape "all goes under macadam."[156]

The building of cottages, motels, restaurants, and commercial centers rose dramatically following the war. The Cape's population grew from thirty-seven thousand in 1940 to forty-seven thousand in 1950. And the summer population exploded. Restaurateurs, motel and guesthouse owners, those renting cottages, and merchants were overjoyed at this new wave of customers. But it also became increasingly obvious that the Cape's transportation infrastructure was not up to the load. Traffic congestion, delays, and exhaust began to plague the north shore along the Old King's Highway, but especially the south shore along Route 28. With the approach of 1950 it was clear a new highway was needed on the Cape.

By the end of the 1940s the modern Cape had emerged. Shifts did occur. Some areas once dominated by small cottages now had those cottages pulled down for far more elaborate ones, while other areas once dominated by prestigious estates fell into decline. The pattern for the Cape's modern condition was set in place. Certain areas, especially along the western south shore and around

Chatham, continued to be dominated by wealthy estates. From Hyannis to Harwich and along the bay from Orleans to Provincetown small cottages and rental units filled in more and more land. Although both Falmouth and Orleans were developing into important commercial centers, Hyannis established its dominant place as *the* commercial center for the Cape in the early years of the twentieth century and continued to hold that position. And the Cape continued to be a place of towns. Falmouth, with its famous Woods Hole institutions, was fixed as the scientific and intellectual center it had become in the nineteenth century. The north shore, which was famous for its sleepy, quaint villages, was by 1940 attracting retirees. As the asparagus and truck farming declined in importance, Eastham became a town of small cottages and permanent homes for those who found work on the Cape year around.[157] Wellfleet's town center, good harbor, and world famous oysters drew to it not only oystermen and boatmen, but also—due to the influence of a few charismatic individuals such as Katharine Dos Passos, Mary McCarthy, Edmund Wilson, and Francis Biddle and his wife Katherine Garrison Chapin Biddle—intellectuals and scholars.

Of all the Cape towns, Provincetown developed the most distinctive character. As the fishing industry collapsed across the Cape in the late nineteenth century, Provincetown fishermen, mostly Portuguese, had continued to work. The town's harbor was deep enough for its fishermen to switch to the bigger beam trawlers, although the diehard line fishermen could not believe anyone would buy fish so badly mangled by the dragnets. To this community of fishermen came artists in the early years of the century, drawn to the area by the special light typical of a northern sandy peninsula as well as relatively cheap housing since housing and rental prices fell from 1890 through 1920.[158] By the 1910s there were art schools attracting students from across the northeast. In 1914 the Provincetown Art Association was founded. The association purchased its own home on Main Street for an arts center and put on the popular Artists Costume Ball. In 1907 Mary Heaton Vorse settled in Provincetown and began publicizing it among writers. By midcentury Provincetown had a strong bohemian community coexisting uncomfortably with a community of Portuguese fishermen and their families. The Cape itself was a rapidly developing population center sitting upon a very fragile and constantly changing sand spit. Yet despite all the changes of those last fifty years, there was little or no awareness of how these changes prefigured the future.

7
THE GOLDEN AGE OF TOURISM

In the second half of the twentieth century, Cape Cod residents and tourists alike carried with them their own individual images of the place. For some, that image entailed beaches, beach umbrellas, beach balls, and mini-golf. It involved outlet stores, bars with names harkening to the age of fishing and whaling, and restaurants with fake buoys and fishnets stretched across their facades. Others carried the image of the Cape as a place of work. Still others imaged the Cape as a collection of small New England villages centered on the church, library, and town hall, a place of refuge and peace. Increasingly, tensions among these different visions were the fodder of local and regional political conflict. Although many thought of the Cape as an exceptional place, its history was part of the larger history of the region and nation. In the postwar years that was a history of growth.

America, at least white America, emerged from the trauma of World War II into an era of prosperity. Never before had so many Americans had so much. For those who stayed on the home front, wartime demand had meant steady employment at good wages, while rationing and shortages of consumer goods encouraged savings. For those who went to war and survived, peace meant government support for higher education with the GI Bill, improved training and skills, and higher salaries. Newly expanded unions meant good wages and job security for more than a third of the work force. The forty-hour workweek and paid vacations provided more Americans than ever with leisure time for recreational pursuits.[1]

Postwar prosperity also led to a dramatic rise in home ownership as suburbs pushed out farms. New highways and increased automobile ownership allowed commuters to live fifteen to twenty miles from the city center. New techniques of home construction and simple septic systems, much of it learned hastily in the rush to build housing for war workers, encouraged the mass building of one- or two-story single-family homes. FHA loans guaranteed builders that there would be buyers for these new homes.[2] By 1950 more Americans lived in suburbs than elsewhere.[3]

Commuters were not the only ones leaving the congestion of the streetcar city. Companies seeking the efficiency of expanding, single-story manufacturing buildings took advantage of cheaper suburban land to build their new factories. Following their customers, retail establishments and eateries also moved beyond the city, often locating themselves in the new suburban shopping malls. Other segments of suburban America increased in number and grew larger, not just middle-class bedroom communities. Employment led to more working-class communities and ethnic enclaves; ethnically diverse suburbs were also to be found outside the urban core.

As America emerged into its age of prosperity, the situation was not quite so rosy for New England. The region's keystone industries, notably textiles, shoes, and paper, had never recovered from the depression of 1920–1921, when capital, factories, and orders went to other parts of the country. War production provided an economic bump, but the decline of orders and closing of factories soon resumed. The benefits of the postwar years did not bypass New England entirely, but were more limited than in other parts of the country. The sluggishness of the postwar New England economy held all but the wealthiest of its residents to more modest lives than their peers in more prosperous sections, which increased pressure for expanded employment. But New England's economic doldrums did not last forever. Those who had made wealth from the old New England industries remained attached to the region and continued to support an intellectual infrastructure which helped lay the foundation for new industries and economic innovations.[4] By the end of the twentieth century those industries and innovations would bring a new prosperity to the region. That prosperity had myriad effects on the region and on the Cape.

Despite its initial postwar depressed condition, New England shared the surge of suburbanization that took over the nation in the years after 1945. Boston had already spun a web out across New England beginning early in the nineteenth century. Banks, railroads, textile centers, universities, and medical centers had linked the region, including Cape Cod, to the city. The post–World War II period represented a filling in of that web with new housing and new manufacturing and retail establishments.

Although the Cape was linked to Boston, it experienced the postwar suburbanization in its own unique fashion. Barnstable County, like much of the rest of the region, found itself in the midst of a home-building boom in the postwar years. These were not homes for commuters, but for retirees and those looking for work in the expanding tourist industry. Urban Boston was too far a commute for those living on the Cape. Only at the beginning of the twenty-first century did the Cape feel the influence of home construction for those who worked off-Cape, and that was mostly confined to the Upper Cape towns of Sandwich and Bourne.[5]

Although commuters did not tie the Cape to the rest of the region, the Cape was tied nonetheless. As the century progressed, the links between the Cape and outside world became more complex and intense as people, capital, and institutions from other areas came to play ever-greater roles in the lives of Cape Codders. Different groups brought to the Cape different assumptions about beauty, esthetics, and other qualities that made a place good to visit or to live in.[6]

Wealthy New Englanders continued to be attached to the Cape for recreation. A new generation of working-class New Englanders, many of whom had used trolleys to go to Nantasket or Revere Beach, now found that they could afford a car and had the time and resources to drive to the beach for a weeklong holiday. People in white-collar professions could now join those who always had the resources to take their leisure for more than a day and farther than a trolley ride. Folks outside of New England took advantage of their newfound prosperity, too, and planned vacations on Cape Cod. All of these tourists brought notions about what a vacation spot should be to places on the Cape where locals already had their own notions.

As the Cape entered into the second half of the twentieth century, these complementary and contradictory assumptions were acted out in the physical place we know as Cape Cod, a thin strip of sand with a single freshwater aquifer, buffeted about by wind and waves. Many of the new visitors to the Cape brought a history and childhood memory of vacations that involved beach crowds, food shacks, arcades, and souvenir shops. Others came with memories of hikes in the country and naturalist retreats. Still others brought different conceptions of what a visit or life on the Cape should entail. Others found work catering to and constructing for these varied populations and visions. The politics of much of the postwar era involved the collision that emerged from these differences. It also brought to the Cape a return to the politics of property.

Land ownership and use, always subject to political and societal pressure, once again became a hotbed for local debate. The Cape had never experienced such intense conflicts, at least since the battles over common lands and common rights in the seventeenth and eighteenth centuries. Local battles, at times intertwined with state and national interests and agencies, embroiled Cape communities. Decisions by individuals about how to use land, where to build, what to build, how to live, and how to make a living collectively worked to transform the Cape. The beaches, the fields, the marshes, the forests, the ponds, the rivers and streams, and the aquifer all changed in nature, becoming different from what they had seemed in the past.

That transformation also gave rise to concerns about the effects and consequences of those decisions. Concern became tied first to conservation and then to environmentalism. The processes of air and water flow, plant life, ponds, and ocean edges took on new meanings. That which had been traditional, private, and accepted became ever changing, public, and political. The postwar boom years of the Cape became political years infused with lusty economic growth and a new awareness of environmental issues.

The war years not only brought military personnel and those who worked at the new bases to the Cape, it also engendered a new optimism. Even with gas rationing limiting the ability of visitors to drive to Cape Cod, they kept coming. By 1946 the stream of visitors turned to a river. With the end of rationing and the availability of financial resources saved up during the war years, Americans went on a buying spree. Automobile manufacturers quickly converted from jeeps to cars, and they found an eager market for every car they could produce. Peace and the new détente between unions and manufacturers brought the two-week paid vacation to more Americans, and the automobile and job security encouraged them to travel farther away and for longer periods of time. Instead of an afternoon at a Boston city beach, more people could spend weekends and week vacations at the ocean. By 1951 over 200,000 vacationers came to the Cape each year for summer fun. During the summer months the number of people on the Cape increased almost ten times.

It was not only summer tourists who were drawn to the Cape. Many came, not to visit and play, but to live. In 1950 Barnstable County's population reached 46,805 residents and by 1960 it reached 70,286. The lion's share of those who came to settle were retirees, followed by those who left the more depressed areas of the region for the lure of Cape jobs in construction and transportation.[7] In Eastham, at the end of the nineteenth century, only 5 percent found employment in construction or as general laborers whereas a majority of the population still found employment in fishing and farming. By 1960, 25 percent of the town's labor force worked in construction.[8] In 1960 more than five thousand Cape families depended upon the construction industry.[9] Although employment in transportation and the building trades waxed and waned with the state of the economy, the pattern of decline in fishing and agriculture and the rise of employment linked to construction and tourism continued through the end of the century. By 1980, only 1 percent of the county found employment in agriculture and fishing, while 5 percent worked in construction. Growing interest in local produce and the expansion of aquaculture and chartered sport-fishing expeditions increased the proportion employed in farming and fishing to 1.65 percent in 1995. But that figure was still well below construction and transportation, which together accounted for almost one in ten employed Cape Codders.[10]

Retirees moved to the Cape for milder weather, less-stressful living situations, and convenient services. People coming to the Cape for employment looked not only for jobs but also for housing.[11] Homes quickly spread out in all directions, as fallow land became housing developments. In 1940, just before the Cape's dramatic increase in population, 34 percent of its residents were between the ages of twenty and forty-four. In 1950 that age group, that is, those forming families and taking jobs, rose slightly to 35.2 percent.[12] After 1950 the growth of those in this age span slowed as the influx of retirees pushed the age of the average Cape Codder ever higher. By the 1980s it was the group between the ages of forty and fifty that grew the fastest, doubling over the decade. In 1960 the mean age of those who called Cape Cod their home was 30.8 years, by 1970 it was 34.3 years, and by 1990 it reached 39.5 years, making Barnstable the county with the oldest population in the state.[13] Whether people came to the Cape to work, to retire, or to vacation, they needed places to live, to stay, to eat and shop. Meeting these needs became the new engine of growth for the Cape.

Seeing opportunity in this potential flood of summer visitors, developers bought land previously thought unsuited for building, then marked off lots and sold them to eager buyers who built cottages as fast as they could get wood, nails, and construction workers to the building sites. The lots and cottages Joshua Nickerson had been holding since the late 1920s began to sell again, and the remaining lots of Samuel Hannah's Homestead Trust were finally sold off in the postwar period. In 1950 almost sixteen thousand summer homes clustered about the Cape's coastline.[14] Builders and investors constructed more than summer homes. For those visitors looking for only a weekend or a few days at the beach, developers built motels, and by 1957 the Cape supported 125 of them. These numbers continued to rise to 288 by the end of the century.[15]

Because land prices away from the beach were still low through the 1950s, most of these motels were small affairs, often built by local landowners and builders. They consisted of a series of small one-room cottages clustered together alongside the highway, or a strip of single rooms with an office and perhaps a restaurant attached. The grounds around the motels were left in sand, heather, bluestem, and beach grass, mowed once in the spring if at all. The limited service consisted of clean sheets and towels and a quick cleaning of the rooms between tenants. Family members and college and high school students off for the summer provided the labor; owners did most of the repairs.

NEW BUYERS, NEW PRODUCTS, NEW RETAILERS

Since the beginning of the twentieth century, large retailers and chain stores across the country were replacing small mom-and-pop stores as economies of scale pushed the smaller operations out of the market. Because of its depressed condition this trend bypassed the Cape except for a few isolated pockets in Hyannis and Falmouth. But with the dramatic countrywide upturn in tourism during the postwar period, national or Boston-based economic interests began to push out on to the Cape. That process slowly transformed local economies and eventually the local landscape.

For those renting a cottage or staying in a summer home, big-city grocery stores such as A&P and Acme, with their large parking lots and expansive buildings, competed with local farm stands and small country stores.[16] Cape truck farmers slowly felt pressured to abandon or drastically reduce their crops due to competitive high-volume producers offering low prices, the emergence of packaged frozen and prepared foods, and rising wages for farm hands. Dairy farms were the first to go, followed by the larger truck and poultry farmers. Cranberry and smaller truck farmers hung on, but more and more of the food on the tables of those who lived on and visited Cape Cod came from purchases made off-Cape in large chain grocery stores rather than at local farmers' stands.[17]

Furthermore, to reduce costs, developers preferred old pastures for construction sites. Between 1950 and 1970, pastureland decreased 88 percent while tilled land decreased 54 percent. Cropland dropped from 7,771 acres in 1950 to 1,788 in 1970. This trend continued for the rest of the century and into the next, until rising slightly to 2,029 in 2007. Pastureland, which stood at 2,618 acres in 1950, fell to 140 acres in 1970; it also rose again, to 704 acres in 2007. Housing, shopping malls, highways, and airports replaced farms. Between 1950 and 1970 agricultural land and publically owned open space fell from 15 percent of Barnstable County to 8 percent, while residential land went from 5 percent to 14 percent. In the twenty years after 1975, 140 acres of shopping centers were built, most of them occupied by large national or regional firms. Over six thousand acres went to recreation, a third of them for golf courses.[18]

Most of the new development on Cape Cod, particularly along the southern coast in the immediate postwar years, consisted of either small summer cottages or homes for retirees. Open space stretching back from Route 28 from Centerville to Chatham began to fill-in with cottages. Further inland, year-round, single-story "Cape Cod" homes for retirees took over old pastures and wooded areas.[19] The completion of the Mid-Cape Highway from the Route 134 exit to Hyannis led to a new line of commercial development up Highway 134 from Route 6 to Hyannis center, and in 1970 the Cape Cod Mall was built halfway between Hyannis center and the Mid-Cape Highway. These commercial centers dominated land along the highway with their buildings and shops. Parking lots for automobiles dramatically increased the amount of paved-over impervious surfaces. Stretching out in front of, beside, and behind restaurants, motels, stores, and shopping centers, these "wastelands" resulted in polluted water runoff. By the end of the 1950s outlet stores, supermarkets, shops, and miniature golf courses cluttered the visual landscape of the Cape highways. Towns across the Cape implemented zoning and planning boards to try to control this explosive growth in the late 1940s and 1950s, but the pressure for development left many towns feeling they were playing catch-up. Initially mid-Cape towns allowed strip commercial development along Route 28, and the ensuing mass of commercial activity soon created a new image of Cape Cod that was closer to that of the coastal resort communities of New Jersey or southern Long Island than that of the quaint villages depicted in Cape vacation publicity.[20]

The Cape's building boom flowed through cycles linked to larger forces driving the national economy such as monetary policy, interest rates, and consumer spending over the rest of the century, but the trend line was upward. Builders constructed so many motels along Route 28 that in 1984 Yarmouth

Increasingly, Cape Cod vacationers flocked to ocean-side beaches and traveled to them by private automobiles. Photo courtesy of the Library of Congress.

passed a ban on further construction of new motels.[21] As drivers traveled the Cape's roads, trees, farms, and villages disappeared behind fields of parking and walls of stores.

For the ever-growing middle classes of Americans, vacations meant family time and family entertainment. The cottage communities that were spreading out around Route 28 became centers of family entertainment. The whole family would trundle down to the shore loaded down with beach umbrellas, chairs, towels, and blankets. Because of the widespread belief that swimming on a full stomach caused muscle cramps, lunch required leaving the water for at least an hour after eating. Families returned to the cottage for the midday meal and afterward either headed to the beach or went on to the entertainment and shopping areas that stretched out along Route 28.

When golf became popular among the American upper classes, resorts catering to the middle class began building small putting greens. By the 1920s, very short putting lanes on artificial turf replaced these putting greens. These golf-putting courses went out of favor in the Depression only to be brought back as increasingly elaborate miniature golf courses in the 1940s and 1950s. Mini-golf courses with their twists, bends, and tunnels became identified with coastal resort communities and popular on the Cape. Unlike golf, mini-golf did not require acres and acres of well-tended lawns and greens. A nine- or eighteen-hole mini-golf course could be set up on a single acre. And mini-golf was fun for the whole family. Although the obstacles in these courses reflected such traditional Cape themes as fishing boats, windmills, and lobster traps, there was nothing traditional about how they used the land. Mini-golf courses, along with strip malls, souvenir shops, and restaurants with sea themes and faux seashore props—from nets, buoys, pier pillars, and pirate flags to plastic fish—became the new visual of Cape Cod. And for many visitors this image was the very reason they made the trek to the Cape.

DEMOGRAPHIC GROWTH

The Cape's beauty, its relatively mild weather, and the job opportunities in construction and service continued to draw ever-more people including a growing number of Boston commuters. With the exception of Provincetown, which lost population between 1950 and 1970 due to the decline in the fishing industry, towns across the Cape grew in population. Between 1950 and 1970 Barnstable County's population grew by 106 percent, well above the commonwealth's overall increase of 21 percent. (Over the last half of the century it grew by 400 percent.)

Almost a hundred thousand people called the Cape their permanent home in 1970, but this dramatic growth was dwarfed by population growth during the 1970s. Over 93 percent of the population increase was due to in-migration, consisting mostly of retirees and those looking for work. Within six years of the 1970 census, the Cape added another 31,193 residents (33 percent). Chatham and Orleans each added over 150 percent to its population over the twenty-five years after 1950, Wellfleet grew by 78 percent, and Eastham tripled its population. Between 1970 and 1975, the Cape added 21,074 dwelling units, an addition of over 31 percent.[22] By 1976 the twenty-year period of population decline in Provincetown had ended, and by 1980, the town gained 21.5 percent in population. During the same period, Truro added 20.4 percent more permanent residents, and Wellfleet added 26.7 percent, Eastham grew by 69.9 percent, and Orleans added 73.7 percent more residents. Brewster's increase was the most significant with a 192 percent increase. Mashpee followed with a 187.3 percent increase. Dennis added 91.5 percent to its population, and Sandwich added 66.6 percent. The town of Barnstable grew 55.7 percent over the decade while Yarmouth added 53.3 percent to its population. Harwich added 52.3 percent and Falmouth added 48.3 percent. Chatham gained 33.3 percent in population and Bourne added 9.8 percent. In Eastham wholesale and retail employment accounted for 43 percent of employment whereas construction accounted for almost 25 percent. In 1970, 25 percent of the Cape's population was composed of retirees, and almost 17 percent of Cape residents were over 65, compared to the 12.2 percent it was in 1940.[23] By the end of the century almost a quarter of the population was older than sixty-five.[24]

In 1974, over 6.25 million tourist dollars were spent on the Cape and the trend seemed ever upward.[25] The number of vacation homes on the Cape jumped from 15,782 in 1950 to over 56,000 by 1980.[26] Summer-home owners and retirees accounted for half of the Cape's economy by the 1980s, and summer visitors accounting for another quarter.[27] The boom in building slowed in the late 1980s as a result of recessions in the national economy in the early 1980s and the technology recession in the late 1980s, which hit the region particularly hard. With the recovery in the 1990s, the building surged again and the numbers of retail establishments and restaurants grew along with the growth of vacation homes, rental cottages, motels, and parking lots.[28]

During the summer months, the population swelled to over four hundred thousand. Eastham's 1979 summer population was 400 percent higher than its winter population whereas Wellfleet's was 800 percent higher. Most of the development on the Cape in the period before 1970 was on the Cape's south shore, but cottages also sprang up across the Outer Cape. Along Pleasant Bay in Orleans, where there were only 150 shorefront houses and dozens of summer camps in 1944, hundreds of new and increasingly larger homes filled in the landscape. The dramatic increase in land values and taxes pushed out the summer camps with the last ones shutting down in the 1980s. Camp land, with its valuable waterfront, went up for sale at prices that discouraged the small cottages typical of earlier summer-home construction.[29]

On lots selling for prices higher than suburban lots near Boston, homeowners built houses that began to compete with their luxurious suburban counterparts. Land around Orleans filled in with houses, and drew commercial activity rivaling Hyannis. Besides the Pleasant Bay area, new homes were going up along Cape Cod Bay from Sandwich to Truro. As in the case of Pleasant Bay, the increased cost of land encouraged the building of more expensive homes.

CARS RESHAPE THE CAPE

If the automobile had become a more important component of tourism on Cape Cod after 1920, by the 1950s it, along with limited bus and ferry service, became the only means of vacationing there. Train service east of Yarmouth ended in 1935. The service to Yarmouth and towns to the west, such as Falmouth and Barnstable, suffered falling ridership following the war. This led to the elimination of all but summer weekend runs, which ended in 1964 even as car ridership increased. Traffic followed development. The heaviest building went on along Route 28 on the southern Cape, which connected to the canal bridge at Bourne and handled cars coming up from New York, Connecticut, Rhode Island, and southern Massachusetts. By 1948 congestion on Route 28 overwhelmed the highway's capacity. Route 6 (now Route 6A), which intersected with the Sagamore Bridge and handled cars coming from the Boston area and north, was not in much better shape, with weekend stop-and-go traffic and abundant car exhaust.

In 1949 a period of highway catch-up began first with a new Mid-Cape Highway, Route 6, along the moraine that ran down the center of the Cape. But the completion of this new, limited-access Route 6, extending from Sandwich to the Hyannis turnoff in Barnstable, did not end the problems of clogged Cape highways. More and better construction only encouraged more congestion on its roadways. In 1954 the new Mid-Cape Highway was widened to four lanes to accommodate the increased traffic, and a two-lane extension was added from Barnstable to Orleans. In 1955, a four-lane highway between Truro and Provincetown was built, and in 1959 a four-lane highway stretched from the Bourne Bridge to Falmouth. By 1970 highways and airports gobbled up over twenty-eight thousand acres of land. Increased highway access encouraged more visitors coming to the Cape hunting for accommodations and recreation.[30]

Except for those who arrived in Provincetown by ferry or those traveling by bus, Cape Cod vacationers were automobile-bound. Once on the Cape they used their cars to shuttle families and beach gear to any number of beaches or ponds. They motored on to grocery stores and restaurants. Rainy days found them driving to the nearest gallery, shop, movie theater, or mall. Each location accommodated the automobile with an expansive parking lot, yet demand for parking soon outstripped the lots. Already by the 1970s, it was noted that parking lots for popular beaches were filled and turning away cars.[31] By the end of the century, Cape radio stations were reporting not only on the weather and water temperature but also on which beach parking lots were full, and signs along Routes 6 informed drivers which if any beaches still had available parking spaces.

Although many of the new restaurants, motels, and stores attempted to present themselves as part of the Cape by designing or embellishing their buildings in retro Cape style—the Christmas Tree Store in Sandwich added a giant thatched roof and attached an imitation Cape windmill to the building—such details (never mind the ubiquitous seafaring props) failed to bring back the past, and little could be done to modify large expanses of asphalt-covered parking lots.

Although the transformation of the traditional Cape appearance reflected a growing prosperity and a source of wealth for many, it also created unease for others. Tensions began to emerge between those who prospered, or hoped to do so, by the new links to the broader economy and those who feared that the special qualities of Cape Cod would be lost with the increased influence of outside interests. This tension found its first expression in concerns over the changed vistas of Cape Cod.

CONSERVATION VERSUS DEVELOPMENT

The post–World War II Cape Cod population was a mix of old Cape families, new retirees, and those who moved to the Cape to take advantage of job opportunities. Summer residents, many of whom now owned property on the Cape, also claimed an interest. These groups thought differently about what the Cape meant for them and their future; the individuals within these groups had their own unique concerns as well as different visions of the future. For some, expanding commercial centers and housing developments meant jobs and opportunity for themselves and their families. It meant rising property values and increased property taxes to support school expansion. It meant convenience and greater comfort. For others it meant higher prices for unused land, which in turn meant greater wealth. For still others it meant wealthy customers willing to pay high prices for goods and services and higher labor costs. Yet that opportunity also entailed a loss. Increased land and housing prices potentially closed the possibility of future generations being able to afford to live on the Cape.[32] For other Cape Codders the increased development represented a loss of a traditional community and the loss of a wild space for walks and hikes. It meant spoiled vistas, crowds and traffic congestion, polluted ponds, and crowded beaches.

Along the coast, motels, restaurants, strip malls, parking lots, and cottages grew faster than trees. To many Cape Codders this growth seemed to overwhelm the Cape that they knew and loved. Led by some old Cape families, the towns of Eastham, Dennis, Provincetown, Yarmouth, and Bourne in 1950 passed restrictive land-zoning ordinances to control the building boom that swept over them in the postwar years. Barnstable faced dramatic growth of its commercial center at Hyannis, and Falmouth faced significant uncontrolled growth even before the postwar boom. Both towns already had zoning on the books. Their zoning legislation focused on limiting the development along the Cape highways, which was seen as undermining the traditional character of the Cape—or at least as that traditional character had come to be idealized.

In 1954 Norman Cook, the secretary of the Cape Cod Chamber of Commerce, argued that strip development was destroying old Cape towns and would ultimately destroy the Cape's tourist cachet. He urged the chamber to lobby for state intervention to limit development on the Cape. In response to Norman Cook's recommendation, the commonwealth's Department of Commerce issued a report in 1955 stating that development on the Cape was becoming a problem and urging the creation of a regional planning district. The Cape Cod Chamber of Commerce joined the call for the towns to form such a regional body.[33] Although the initial argument for the zoning was to limit the proliferation of strip development, regulations on housing lot sizes soon followed.

The requirement that proposed zoning laws pass by a majority at town meetings, combined with the opposition to zoning restrictions from builders and developers, limited the reach of the zoning ordinances that did get passed. Despite these weak ordinances, development on the Cape continued seemingly without any restraint. Population jumped from 46,805 in 1950 to 70,350 in 1960 and 96,656 in 1970. The pace picked up in the 1970s, when the population grew by 53 percent to 147,925 in 1980, and it gained another 26 percent to reach 186,605 by 1990. Although the growth slowed in the last decade of the century, 222,230 people called the Cape home in 2000.[34] In 1963 a state report warned that the Cape had taken on "a crowded, mass-produced atmosphere" that would only get worse.[35] Two years later, Cape voters supported the creation of the Cape

Cod Planning and Economic Development Commission to advise the towns on development issues.

Concerned that increased traffic would overwhelm their towns, residents of the Outer Cape fought against widening the Mid-Cape Highway from Dennis to Orleans to four lanes, leaving intact the famous thirteen miles of so-called Suicide Alley. In 1968 again under the leadership of well-established Cape families such as the Nickersons and Whitlocks, Cape Codders came together to form the Association for the Preservation of Cape Cod. The APCC initially lobbied towns to pass zoning restrictions, create historic districts, and set aside preserved land.[36] The APCC was founded, as James Nickerson put it, because "the natural endowments of Cape Cod, its coastal waters, its beaches, its rivers, its rural open spaces, its marshlands, its abundance and purity of freshwater ponds and lakes are fragile. Their span of endurance is short against the abuses and chaotic development that has overwhelmed so many areas."[37]

Not all Cape Codders saw the world as James Nickerson did. The boom in construction and tourism meant jobs and an increase in income for those living on the Cape. The jobs went both to the sons and daughters of old-time Cape Cod families and to newcomers to the Cape. For landowners, many who were also from old Cape families, the boom offered the opportunity to finally realize a significant profit from land they had wrestled with for generations. An acre of land along the water that sold for five hundred dollars in 1940 brought in fifty thousand dollars in 1975 and reached a half-million dollars by the end of the century.[38] The concern about uncontrolled development did not resonate with those looking for work or hoping to sell land in the new tourist economy. Land-use planning and restrictions on building were not high priorities for these people. The conflicts that emerged were often depicted as existing between traditional Cape Codders, who were suspicious of government action that would limit their economic opportunity, and newcomers, particularly retirees, who having gained their "piece of heaven" wanted to protect it from future degradation.[39] As in all things this dichotomy was too simple. Many old Cape Cod families, such as the Nickersons (who themselves were involved with tourism), led the fight to control development, while many of those opposing restraints on development were newcomers anxious to gain from the economic boom that swept the Cape in the postwar years.

Pressure from APCC, other conservation organizations, and the Chamber of Commerce encouraged the creation of historic districts in Yarmouth Port and Sandwich Village along the renamed Route 6A, the highway running

The Golden Age of Tourism

along the Bay side from Sagamore to Orleans. Strong zoning also helped protect the Mid-Cape's north shore from some of the dramatic commercial development of the south shore. As early as 1967 the citizens of Dennis, having witnessed significant loss of marshland, spent $625,000 to buy 1,400 acres of wetlands, or 11 percent of the town's area, in order to protect them from development.[40]

Concerned about protecting water supplies from sewage seepage, Upper and Mid-Cape towns, under the leadership of their conservation commissions and "Open Space" committees, began buying land to hold as undeveloped.[41] Nickerson State Park, established in 1934, also on the north shore, protected a huge expanse of land, as did the 83,000-acre Shawme-Crowell State Forest, established in 1923 in South Sandwich. In 1958 the Massachusetts Audubon Society purchased the 366-acre Austin Ornithological Station, which occupied an abandoned asparagus farm, and created a wildlife sanctuary. At the northern tip of the Cape near Provincetown, the commonwealth owned a large tract of land, the Province Lands, inherited from the colonial government. Despite this conservation activity, it was clear that developers had transformed most of the Cape's quaint villages south of the Mid-Cape Highway, from the canal to Orleans, into a vast suburban commercial-residential strip. The pressure to push that development north up to Provincetown was building fast. Concern rose among preservationists, conservationists, new retirees to the Cape, and some Cape visitors: if nothing were done to protect it, the whole Cape would come to resemble the more densely settled commercial strips along Route 28. The fear was that Cape Cod and the Commonwealth of Massachusetts would lose an important iconic tourist attraction.

THE CAPE COD NATIONAL SEASHORE

Conflicts over the different visions of the Cape—a Cape of quaint villages, open vistas, and the vast wild ocean shore, or a Cape of opportunity and employment—expressed themselves in multiple ways: conflict between the purer past and the corrupted modern, conflict between private property and public regulation, and conflict between locals and newcomers. For most Cape Codders and those who came to visit, the central conflict was over land and access to it.

Concern for the loss of open land to development led some towns, such as

Dennis, to appropriate public funds to buy land for conservation. But many conservationists came to believe that the pressure for development and the cost of land purchase was too strong to be overcome by local action. In 1937 the *Cape Cod Beacon*, looking to the model of parks in the western United States, called for a national park that "would prevent further commercialization of dollar-seeking individuals."[42] In 1939 the United States National Park Service, following up on a Civilian Conservation Corps study, proposed a seashore park with the Cape as a possible site. Then, in the mid-1950s, the park service listed "Great Beach, Cape Cod" as its first priority for a new national park.[43]

By the late 1950s a push began, both by the National Park Service and well-connected state politicians. These included Francis Sargent, a future Massachusetts governor and summer resident of Orleans; Senator John F. Kennedy, whose family owned an estate in Hyannis Port; and Senator Leverett Saltonstall —the three men worked together to get a feasibility study for a national park along the Outer Cape. The park service's 1956 study raised a concern for "the vanishing shoreline," noting that a "surging tide of modern development has rolled over vast areas of our national pristine coastal country, wiping out one after another, the natural open spaces. . . . This is happening to Cape Cod. . . . Great Beach is vanishing under buildings. It is time to set aside, preserve, and protect the last of the 'old' Cape."[44] The "Great Beach," or as locals called it, the "back-side" of the Cape, had been spared most of the development which swept across the southern Cape and the western side of the Outer Cape. With some exceptions, such as the development of Bolton Beach in Truro, the rough and cold water, lack of harbors (excepting Salt Pond Bay in Orleans), and high, difficult-to-negotiate dunes discouraged development. Henry Beston's description of the Great Beach in his classic *The Outermost House* differed little from Henry David Thoreau's descriptions seventy-five years earlier.[45]

The commonwealth's two U.S. senators, along with its two most prominent congressmen, Thomas "Tip" O'Neill and Edward Boland, and the Cape's representatives in the state legislature, supported the National Park Service plan. On the other side, the congressman who represented the Cape, Donald Nicholson, opposed the plan, as did his successor in Congress, Hastings Keith. In the late winter of 1959, the staffs of Kennedy and Saltonstall worked on developing a park plan that would be acceptable to Cape property owners and would meet the park service goal of protecting the Outer Cape from continued construction. The plan allowed private property owners within the park to continue to hold their property, but most of the open undeveloped land would be bought by the park and added to already-held government land, such as the military bases in Truro and Wellfleet. It also allowed towns that lay partly within the park to retain authority over town beaches. Conrad Wirth, the director of the National Park Service, came to Eastham in March 1959 to announce the National Seashore Park plan to a crowded town hall meeting. To his surprise many Outer Cape residents opposed the plan. Reverend Earl Luscombe, Wellfleet's Methodist minister, urged Cape Codders to "meet the foe" of park supporters.

Charles Frazier was an influential Wellfleet town selectman with close ties to development interests. In fact, he had played a major role in the rebuilding of the Herring River dike and the cut-off of salt water above the dike in contradiction to the initial agreement. He organized a committee in 1959 to limit the park to only twelve thousand acres. Many residents of the Outer Cape towns resented that some of the land being proposed for the park was town land. They did not want it taken out of their control by an outside agency. Others objected to the park because it took taxable land off the town books. The towns also expressed concern that they would lose control over beaches that brought tourists to the town and revenue to town coffers. Although the park service bought any private land that it took, some landowners and developers, looking at the skyrocketing land prices in the towns of Barnstable, Yarmouth, and Dennis, felt the park was cutting off their opportunity for making a handsome future profit.

Many local families had used the open land on the Outer Cape as a commons from which they hunted, fished, and collected wild berries, mushrooms, and bayberry leaves to sell to local candle makers. They raised concerns that the National Park Service would cut off their access to these common resources. Of course continued development would have meant the elimination of the use of this land as a commons, but the combination of limits on future economic gain and a feeling that the park was being imposed on the local communities by an outside force fueled opposition.

Despite this opposition, the compromise plan drawn up by Kennedy and Saltonstall won enough support on the Cape and elsewhere to pass Congress, and Public Law 87-126 authorized the establishment of the Cape Cod National Seashore on August 7, 1961.[46] Still, opposition to the park did not end with its creation. The Cape Cod National Seashore created an advisory board made up of local representatives, and four of its members openly opposed the park. Joshua Nickerson II, who was not a direct relation to the James Nickerson family of conservationists, lobbied the representatives from other towns to sabotage the Park, but to no avail.[47] Once created, the park encompassed some 27,700 acres of land and 17,000 acres of marsh and water.[48] The park included the Great Beach from Orleans to Provincetown. It took over the Province Lands in Provincetown (which had previously been in state hands), Pilgrim Springs State Park in Truro, Great Island in Wellfleet, and Nauset Marsh and Fort Hill in Eastham. In Eastham and South Wellfleet, the park controlled the Great Beach as well as upland open plains, red maple and white cedar swamps, saltwater and freshwater marshland, and significant sections of reforested woodlands. In the Province Lands, the park owned vast stretches of open dunes, a beech forest, and wetland marshes.

Management of these diverse ecosystems required decisions not only about how to control development, but also what kind of "natural beauty" and "environmental integrity" the park should protect. Without intervention, much of the open plain of Eastham and Wellfleet would return to forest, leading to habitat reduction for ground nesting birds and other fauna as well as the loss of grassland flora. The protection of the historic Cape, part of the park's mission, involved the park in decisions about what was the "historic Cape." For many visitors and residents that meant keeping open vistas, fields, and meadows. For this purpose the National Park Service practiced controlled burns and some mowing, but in doing so it reduced the natural reforestation that was occurring across the Cape's open land.

Because of compromises needed to get the bill through Congress, the Cape Cod National Seashore did not hold all the land within its boundaries. Within the park were over 550 privately owned homes whose owners retained rights to their home, some for life tenancy and others with the right to own and sell. Yet the existence of the park did limit the construction and development that was sweeping eastward along the Cape.

LOCALS AND RETIREES WEIGH IN: "OLD CAPE" VERSUS DEVELOPMENT

The Cape Cod National Seashore brought attention to the issue of uncontrolled development and provided protection for a significant portion of the Outer Cape, but it did not put an end to land management problems. Demand for land and the opportunity for profit put pressure on open land from the canal

The Golden Age of Tourism 167

to Orleans. North of Orleans, houses, shopping areas, motels, restaurants, and mini-golf courses were quickly filling in land outside the park boundary. One consequence of the park taking over twenty thousand acres of land was that the value of the remaining acres increased dramatically. The economics of increased land values drove up the cost of the development on that land, whether larger stores or more expensive homes. For those with land to sell or skills to be put to use developing, designing, or building on the land, this meant jobs and profits. For those wanting to protect an image of what was imagined as the traditional Cape and its beautiful vistas, this development was a threat.

In town meetings across the Cape tempers exploded over the ways in which controlling development to protect open spaces, scenic views, and delicate landscapes could result in lost jobs and opportunity. Concerned that the Cape's growing population would threaten water supplies and that existing solid waste management plans of the various Cape towns were inadequate for projected population growth, Barnstable County established the Cape Cod Planning and Economic Development Commission (CCPEDC) in 1965. (It was the predecessor of the Cape Cod Commission, established in 1987.) The CCPEDC attempted to establish water protection bylaws and solid waste management agreements, but it was limited in its ability to entice individual towns into cooperative action. Jealously defending local autonomy, Cape towns resisted giving up power to a regional authority or submitting to a region-wide agreement. Each town perceived its situation as unique, and each worried it would end up bearing the burden of another town's problems.[49]

The rapid rates of development in the 1970s and the mid-1980s focused attention on the negative impact that development could have on the whole Cape experience.[50] In 1986 and 1987 the CCPEDC directed "Prospect: Cape Cod," a nine-month strategic planning initiative, which brought together diverse members of the Cape Cod community to discuss the problems that dramatic population growth would pose for the Cape's future. "Prospect: Cape Cod" had an open-ended agenda, and although the CCPEDC initially focused on the problems of waste management and safe water, most of the public attention was directed at issues in plain sight: too many houses spoiling the landscape or ugly parking lots and strip malls destroying the quaint Cape communities.[51] One of the major proposals put forward by the initiative was to create a regional land-use agency to address problems in the decision making that affected local land use.[52]

CAMPAIGN FOR LIMITS ON DEVELOPMENT

Cars and traffic proved to be powerful factors in the politics of the Cape's development. Highway construction did not relieve the traffic problems associated with a Cape vacation. By 1977, despite the building of a new expressway from Boston to the Sagamore Bridge and the Mid-Cape Highway, visitors faced potentially long delays (especially on weekends) getting to their Cape vacation spots.[53] Once on the Cape, visitors had difficulty finding parking. Lots close to popular beaches filled early in the day. Bumper-to-bumper traffic on the Cape made short trips to and from the beach or the grocery store increasingly difficult and time consuming.[54]

Faced with rising traffic congestion, unsightly development, and diminished open space, Cape Codders began to grapple with the idea of limiting development. In 1975 the *National Geographic* commented on the increasingly negative impact of tourism on the Cape, referring to Hyannis as "a circus of shopping centers, motels and billboards."[55] Norman Cook, the director of the CCPEDC in the 1970s, called for "a construction slowdown" because the Cape did "not have the water or land to support many more people."[56] By 1980, however, the Cape was in the midst of a building boom with several new malls, resort and retirement communities, and gated developments in Brewster, Mashpee, Falmouth, and Yarmouth Port joining those in Hyannis and Orleans, and the boom continued into the mid 1980s. In 1987 the CCPEDC, now under the leadership of Armando Carbonell, followed up on the "Prospect: Cape Cod" recommendations by calling for the creation of a Cape Cod Commission to take control of growth.[57] The need for a more comprehensive approach became apparent, and the tactics varied. U.S. senator Paul Tsongas called for a moratorium on new construction. Concerned that Tsongas's plea would lead to legislation, local citizens and some local business people favored Armando Carbonell's recommendation to head off more restrictive legislation with a Cape Cod Commission (CCC) aimed to control growth. "Prospect: Cape Cod" focused on the need for a regional land use agency that could address the weakness of local authorities when up against the legal and political influence of large developers. It stressed that the impact of these large developments often stretched across town boundaries.[58] It was also noted that, although the state had procedures governing the permits issued for large developments, most development on the Cape, albeit large by Cape standards, did not trigger state intervention.

Following the ideas outlined in "Prospect: Cape Cod" and modeling on the Martha's Vineyard Commission, Donald Connors, a local attorney who had worked with the CCPEDC to put together "Prospect: Cape Cod," drafted legislation to create a regional land use planning and regulatory agency.

Once the campaign for the CCC was initiated, many in the business community backed away and rallied the building and real estate community to organize against it. The development industry poured money and lobbying efforts into defeating the CCC, but in 1988 over three-quarters of Cape voters (76 percent) endorsed the creation of the commission, leading to the signing of the Cape Cod Commission Act in January of 1990 by Governor Michael S. Dukakis. The act required an additional countywide ratification vote. Again the business and development interests lobbied hard against the act, but in March of 1990, 53 percent of Cape Codders voted for it.[59]

In 1991 the Cape Cod Commission conducted a survey of residents on the future of the region and found that 75 percent rejected further expansion of the tourist industry, particularly factory outlets, restaurants, hotels and motels, and miniature golf.[60] The new commission was empowered to prepare and oversee the implementation of a regional land use policy plan; to recommend areas for designation "as districts of critical planning concern" requiring special protections, and to review and regulate developments of regional impact, particularly development projects of over 10,000 square feet of commercial space or thirty housing units. The commission also required projects to mitigate their impact on groundwater. It published a regional policy plan with guidelines for development.[61] Once established, the Cape Cod Commission won public support with a plan encompassing much of what Cape citizens were telling pollsters and local newspaper reporters they wanted.

Public sentiment, however, did not necessarily drive policy. Under state law, town actions involving zoning changes require a two-thirds majority. The development industry often mobilized against actions that it considered contrary to its interests. Local citizens found it difficult to get their concerns codified into practices that would protect what many came to believe was a fragile and delicate environment. With millions of dollars at stake, those with the resources to hire the best legal talent and the time to pursue their interests had a distinct advantage over residents concerned over the environment.[62] Local citizens could and did mobilize and effected some significant restrictions on Cape development, but the power of money and the market tilted the playing field against them.

By the end of the century, despite all the obstacles, the Cape could look on several environmental accomplishments. The Cape Cod National Seashore protected a large chunk of the Outer Cape from uncontrolled development. Local town open land committees had purchased significant blocks of land. In 1999 the Cape voters adopted the Cape Cod Land Bank, which placed a 3 percent surcharge on local property taxes for acquiring open space. This gave the local open spaces committees a dedicated stream of funds for land preservation.[63] Various town land trust organizations continued to buy development rights and/or land for preservation.[64] At the turn of the twenty-first century these town land trusts came together along with other Cape conservation organizations such as the Association to Preserve Cape Cod and the Massachusetts Audubon Society to form the Compact of Cape Cod Conservation Trusts to coordinate land preservation. Through these combined efforts much Cape land has been protected from development and misuse. The Cape Cod Commission and town leaders renewed efforts to increase lot sizes; it fought for rigorous state requirements for Title V septic system inspection and signoff and better monitoring of wastes. These actions together worked to slow the eutrophication of ponds and help protect wetland water. Developers found they had to be more careful in their plans in order to pass through the many state and local hurdles. For example, community opposition defeated a large Stop and Shop supermarket planned for Truro.

TREES AND GROWTH

Although development and congestion dominated the visual landscape along the major highways, further inland the shift from farming to tourism gave hope that the new economy would put less pressure on the land. Where there had previously been open land, forests began to grow. Large tracts of land that had been cleared for pastures and turnip fields were left fallow. Some of this land fell under the bulldozer and became homes or parking lots for the stores and motels that were lining the highways, but in some other areas bushes and trees began to grow. Even among the new homes filling in the land behind the main highways, trees were planted. In 1950, on 173,894 acres of land, equivalent to almost two-thirds of the Cape, trees were beginning to dominate, and by the end of the 1960s forest covered much of the Cape between the Canal and Truro.[65] Although home construction and shopping and commercial development slowed forest regrowth on the Mid-Cape, the Outer Cape reforested faster than it developed through the end of the century.[66] Nineteenth-century visitors to the Cape marveled at the vast open landscape. Homes a mile in from the coast had dramatic views of the sea. But increasingly forests replaced those open vistas, and water views were lost to trees.

Measured by the increased forest growth since the nineteenth century, the land of the Cape seemed to be restoring, although the forest of the late twentieth century was different from that viewed by the early colonial settlers.[67] Because of the pattern of fires, colonial settlers found mostly pitch pine and scrub oak forests. Pitch pine, which is not shade-tolerant, thrived where fires killed competing tree species. Abandoned fields and pastures also favored pitch pine and scrub oak, and these were the regrowing forests that twentieth-century Cape visitors noticed. Concerned with property and forests, fire departments across the Cape, like their counterparts in the state and national park services, practiced fire suppression. Under such conditions, other trees such as white pine, black locust, white oak, and bur oak pushed into the less shade-tolerant pitch pine and scrub oak forests. The Cape forest at midcentury more closely resembled the colonial Cape forest, but today's forests are more deciduous.

On the Cape one can still catch glimpses in Barnstable and Truro of what most of the Cape was like a hundred years ago. Undergrowth on the forest floor gives some evidence of the previous condition of the land. Where huckleberries grow under the trees of today's Cape there was most likely a woodlot. Where grass grows on the forest floor, there was previously a plowed field.[68]

The growth of the surface forest of the Cape gave weight to an argument that the new economy of tourism was healing the Cape's ecosystem. But development and additional residents and visitors to the Cape impacted more than just the Cape's visual ambiance. By the second half of the twentieth century the Cape had more permanent residents than ever before in its history and summertime densities matched thickly settled suburbs. The number of people who called the Cape their home grew to over 222,000 by the end of the twentieth century. A large percentage of these new residents were retirees. Barnstable County had the seventh-largest percentage of persons over sixty-five in the nation. In addition to retirees, people looking for work in the construction and tourist industries arrived, and for the fifty years after World War II there were plenty of jobs in both, at least during the summer months. While only 7 percent of the Cape was developed land in 1951, over 38 percent was developed by 1997. Homes, restaurants, shopping malls, and motels went up in record speed.

The economics of tourism fed expansion. By the end of the twentieth century, Cape Codders were beginning to realize that the regime of recreation and tourism had substantial costs. With the dramatic increase in land prices, owners had more difficulty leaving land fallow or in traditional use. People who came to the Cape to take advantage of the booming growth in construction needed more homes to build or faced unemployment. Real estate agents wanted more lots and homes to sell, restaurant owners wanted more parking and floor space to accommodate more customers. For many of these people restraints on development represented a threat to their vision of material betterment.[69] The Army Corps of Engineers published a study in 1979 that found that, although many towns on the Cape were aware of the importance of preserving the environment, "many residents are also very concerned that development continues to provide stimulus for the local economies."[70]

Not only did this fragile strip of sand bear a heavier load of humans than ever before, but also this new generation of humans had heavier environmental footsteps than those that preceded them. The environmental footsteps of this new generation of the Cape's visitors and residents were as not fleeting as footsteps in the sand; some had greater impact than others. And although some environmental footsteps were visible to the naked eye, others were invisible.

8
PROBLEMS IN PARADISE

The booming decades after World War II put money in the pockets of Cape developers, construction workers, retailers, restaurateurs, and sellers of land. But that economic prosperity also carried an environmental cost, most of which was invisible. With increased population came increased sewage, increased solid waste, and increased water use. Demand for vacation homes with water views or proximity to shorelines led to filling in of wetlands and to home construction on fragile sand dunes or flood-prone lowlands. More recreation enthusiasts meant more marinas, more dredged harbors, more powerboats, and more golf—and to more battles over land use. If the last part of the twentieth century saw conflict over the esthetics of overdevelopment, it also saw substantial political collision over how to deal with growth's environmental outcomes. These conflicts pitted neighbor against neighbor. They also involved ever-greater intervention by outside forces. The politics of the period were not just the politics of growth and restraint; they were also the politics of science and awareness. As with much of the nation in the years following the war, changes in the physical world of the Cape, engendered by its economic success, brought an increased awareness of the environmental and health costs of that success. That consciousness itself drove and was intertwined with politics on the local, state, and national levels.

DEVELOPMENT, FOR BETTER OR WORSE?

Development changed the visual landscape of the Cape, although Cape Codders disagreed about just how negative these changes might be. Seashell and seafaring motifs on the facades of shops and Cape-themed restaurants, as well as the parking lots to accommodate their patrons, however unsightly, were hardly the worst problems pressing against the Cape. Development increased flood damage from storms, exacerbated the loss of wetlands, and compromised the region's waters. State and federal investigators and agencies, concerned about the costs of disaster relief, pressured local communities to restrict building in sensitive areas. Developers and individuals owning coastal land saw such development as their right as property owners, and they pressured local communities to allow them to use their land as they themselves saw fit. As open land became scarcer, and its value increased, a medley of conflicts heated up: long-term environmental degradation versus short-term financial gains, local versus regional or national concerns, and private versus public interests. Increasingly the politics of the Cape centered on these conflicts.

Wetlands and tidal estuaries had been under pressure since the early nineteenth century. The building of wharves, the dredging of harbor channels, and railroad construction altered or reduced wetlands. But the more massive twentieth-century projects were the ones that dramatically changed the ecology of coastal areas. Marshlands, such as the Herring River in Wellfleet, were diked and drained. Inlets were dredged, walled, and re-formed to create docks and marinas for recreational boating. Builders filled in marshland to create new land for homes and commercial development.[1] The dumping of dredged material also eliminated some marshland.[2] Between the years 1951 and 1972, salt marsh acreage decreased from 16,142 to 13,184 acres. Development and highway construction accounted for most of this reduction.[3]

Filling in of marshland was not the only source of coastal transformation that accelerated in the postwar period. As Rachel Carson noted in *The Edge of the Sea*, the strange and elusive shoreline is a persistent lure for vacationers. For much of the nineteenth century, the Cape's coast was filled with unsavory (though increasingly declining) saltworks, working wharves, and fish processing. The coming of vacationers late in the century made the edge of the sea the most desirable property and a wave of new construction soon filled in beachfronts.

Hoping to encourage the construction of wharves to facilitate trade, the Massachusetts colonial government had modified traditional common law and allowed the ownership of shore property to extend from the high- to the low-water mark. This gave waterfront property owners greater leeway in developing the shoreline. However, the increase in housing and development along the coast not only expanded access to the beach for vacationers, but also increased susceptibility to damage from storms. As Henry David Thoreau wrote, the Cape sticks out into the wild Atlantic like the arm of an athlete "boxing with northeast storms." Winter brings hard cold storms to the Cape that hammer it from the northeast while in the summer, more occasionally, hurricanes sweep up from the south. These storms, both northeasters and hurricanes, push up water in front of them and increase tidal flooding, a concern because 7.75 percent of the Cape is subject to tidal flooding.

Riding the Storms

Between 1901 and 2000, twenty-eight significant hurricanes swept onto the Cape focusing their energy on the southern and western Cape. Five of those hurricanes, the ones in 1938, 1944, 1954, 1960, and 1991, caused the most flooding. On September 21, 1938, the storm that came to be known as the Great New England Hurricane swept up the Atlantic coast at sixty miles per hour. With its eye holding to the warm Gulf Stream waters, it kept its intensity as a Category 3 to a Category 4 hurricane until it hit land along the southern New England coast. It moved onto land just as New England was experiencing exceptionally high tides. Storm surges of eighteen to twenty-five feet crashed across southwestern Cape Cod. It flooded large sections of Falmouth with eight feet of water.

Although the storm of 1938 produced the highest flood waters (followed by the Great Atlantic Hurricane of 1944), it inflicted less property damage on the Cape than Hurricane Carol on August 31, 1954, which caused over six million dollars' worth of damage. The Great New England Hurricane killed 564 people compared to Hurricane Carol's death toll of 60, but Carol had the greater economic impact. In 1960 Hurricane Donna swept across the Cape tearing up trees, ripping off roofs, and flooding homes. Donna caused fewer deaths than Carol, but almost as much economic damage. This increase in damage was directly related to the increased development along the coast that occurred in the postwar period.[4] Luckily the Cape was spared a major hurricane for the next thirty years until Bob moved on shore in 1991. Thousands evacuated the Cape, causing an eleven-mile backup at the Sagamore and Bourne bridges.

Although Bob was only a Category 2 hurricane and caused only one death in the state, land and building development along the coast dramatically increased its economic destruction, causing over a billion dollars in damages to property, most of which was on the Cape.[5]

The High Price of Water Views

The demand for ocean views encouraged the most dangerous construction. Several new developments were built on land that had experienced severe flooding in the storms of 1938, 1944, 1954, and 1960. The economic costs of tidal flooding in the nineteenth century had been limited because the land in the flood-prone zone had little value. The wealthy avoided the lowland beach areas, preferring higher land with more commanding views, whereas old-time Cape Codders built away from the flood zone specifically to avoid damage to their homes. The limited damages on the Cape from the storms of 1938 and 1944 told of the abandonment of old shoreline activities.[6] By 1950, however, vacationers wanted beachfront or water-view property and were willing to pay a high price for it. Builders and buyers were willing to take the risk of flooding for the opportunity of selling or owning this highly valued land.[7]

A developer in East Dennis not only built homes on land which had been flooded in earlier storms, he even leveled protective barrier dunes in front of his new houses in order to give potential owners "ocean views."[8] The sand removed from the barrier dune was deposited as fill in low areas behind the dune to allow for more development.[9] The first floors of this developer's homes were less than four feet above the normal high tide level. If they had been built before the 1944 storm, their first floors would have been almost five feet under water during the storm. Failing to take a lesson from the past, the developer of this project even built homes below the level of an older neighboring house that had been badly flooded by the 1954 storm.[10] In West Dennis, a developer admitted to Donald Crane, an investigator for the Massachusetts Water Resources Commission, that, although he was building homes on land filled in by leveling a barrier dune, it made no difference since other developers had already leveled the surrounding barrier dunes.[11] The West Dennis builder, who claimed to Crane that he told the buyers of the possibility of flooding, initially had problems securing financing for his project because of the larger regional banks' concern about the hazards of building in flood-prone areas. Ultimately, feeling that demand would outweigh the risk, a local bank financed the project.[12]

The Massachusetts Water Resources Commission noted that despite the potential for flooding, the owners of highly valued, low-lying property seemed willing to take the risk. One owner whose home was significantly damaged by flooding in the 1954 storm quickly began rebuilding in exactly the same location knowing that he faced the potential of the same thing happening again. Although owners and builders made claims of a willingness to accept the danger of flooding of high-value shorefront property, Crane noted that in fact these same owners were trying to "push the local, state, or national government to underwrite their risk, or take action to mitigate it after the fact."[13] Given that reality, in 1962 the Water Resources Commission recommended that "special building codes should be adopted for building in the flood plain, [and] a map should be displayed of the flood plain so that future purchasers will be warned of the potential hazards of occupying that land."[14] It was not an especially strong recommendation given the commission's own finding that people seemed willing to take the risk if they had some confidence that the local, state, or federal government would step in to mitigate the danger.[15]

WETLANDS: APPRECIATION OR DEGRADATION?

These concerns were deepened by the commonwealth's growing appreciation of the importance of wetlands, bolstered in turn by the science of ecology, which encouraged biologists to look to the interactions between species, and between species and their environment. In 1962 Massachusetts passed the Jones Act, which prohibited the removal, filling, or dredging of "any bank, flat, marsh, meadow or swamp bordering on coastal waters" without the approval of the board of selectmen, the state Department of Public Works (DPW), and the Director of Marine Fisheries. Yet marshlands continued to be filled in and developed on the Cape with the prevailing attitude being "fill first, file second." No one was required to remove fill and no one was fined or jailed. The Jones Act was followed in 1963 by the Coastal Wetlands Protective Act, and two years later by the Coastal Wetlands Restriction Act. These new laws did limit the filling in of wetlands for development, but the problem of marshland being sacrificed to construction remained. Even after the Coastal Wetlands Act was strengthened by passage of the 1972 Wetlands Protection Act, enforcement remained weak because the act regulated but did not prohibit wetland destruction.

In order to be sensitive to local concerns, final decisions on wetland filling were left to local conservation commissions. In 1957 the state had mandated the creation of conservation commissions as official town bodies. On the Cape these commissions, unpaid positions appointed by town selectmen, were not meant to duplicate the actions of the town planning commission or zoning board. Instead they focused singularly on wetlands. They were the main vehicle for marshland protection, but did little to limit building. Many members of these commissions were conscientious citizens. They were also members of the community whose friends and neighbors had a direct interest in filling in a strip of marshland or dredging for a dock. Local residents interested in development made it a point of offering to serve on the conservation commissions, and selectmen, desperate for volunteers, often accommodated.[16] Pressure to make exceptions and allow filling or dredging increased as less and less land was available, and as waterfront land became the most valuable for development.

Although developers continued to fill and drain wetlands and marshlands into the second half of the twentieth century, by the 1970s public sentiment shifted toward protection.[17] Cape citizens rallied against the road construction project to widen Route 6 between Dennis and Truro because it called for cutting through and filling in several wetlands. Letters and testimony against the projects in both Boston and Washington, DC, put them on hold until they were finally shelved due to a shortage of funds.

In Wellfleet, the townspeople who had resisted Lorenzo Dow Baker's pleas to dike the Herring River did not have the political muscle to stop Baker, but nature eventually began to push back against his dike. After sixty years of the dike and the reduction of tidal inflow to less than a tenth of its traditional levels, the floodgates at the mouth of the Herring River became so corroded that salt water began to flow again into one of the largest wetland estuaries on the Cape. Salt-tolerant *Spartina* grass began replacing cattails and other freshwater vegetation, while oysters and other bivalves appeared above the dike and herring ran once again to upriver ponds. Nature began to slowly restore the historic Herring River, much to the pleasure of many local residents.

The Bridge versus the Dike

Not everyone viewed the recovery of the saltwater marsh as a good thing. The deterioration of the dike undermined the road that crossed the mouth of the Herring River to reach the popular beach areas of Great Island and Duck

Harbor. The town now confronted the choice of building a bridge or replacing the floodgate and repairing the dike. The conservation commissions, shellfish harvesters, sport fishermen, and conservationists favored the bridge. The town's selectmen, led by Charles Fraizer, listened to real estate interests and favored rebuilding the dike to protect the reclaimed land upstream. With the majority of the town opposed to rebuilding the dike, pressure built on the selectmen to choose the bridge option.

At that point the state Department of Public Works, a single-minded, traffic-oriented agency with a history of resisting conservation measures, entered the fray. The DPW's voice carried weight in town affairs, and when the agency sent representatives to the town meeting to testify in favor of the rebuilding option, this was the option that prevailed. Opponents did win a partial victory when the meeting agreed to conditions insisted on by the conservation commission. The rebuilt dike would have to allow the same flow of water up the Herring River after the construction of the new dike and floodgate as before, and would have to allow the free passage of alewives to their spawning ponds. The Department of Natural Resources was to affirm these conditions. By 1974 construction was finished, but not in accordance with the agreement. Almost no tidal salt water made it past the dike. Salt levels approached zero, shellfish died, and alewives were cut off from their upstream migration. Ten Wellfleet residents went to court and eventually got an injunction against the DPW. Public outcry and the court case forced the agency to adjust the water allowed through the floodgate, but even with the adjustments, water levels behind the dike were still a foot below the level originally agreed.[18]

Failure to prevent re-diking of the Herring River did not mute public concern for wetland loss on the Outer Cape. Cape citizens and scientists, particularly John Portnoy of the Cape Cod National Seashore, continued to battle to reclaim the Herring River salt marsh and succeeded in restoring a large salt marsh at Hatches Harbor in the Province Lands. In December 2007, a memorandum of understanding among the National Park Service and the towns of Wellfleet and Truro created the Herring River Restoration Committee for the purpose of preparing an Environmental Impact Statement and Environmental Impact Report for restoring the Herring River Salt Marsh. The following year a group of citizens from Wellfleet and Truro formed the Friends of the Herring River to push forward the process of what might become the largest salt marsh restoration on the East Coast.

Overloads of Petroleum and Pesticides, Nutrients and Nitrogen

But dikes and filling were not the only factors impacting wetlands. Those areas that remained were under increased pressure from pollution. Boats at marinas, docks, and harbor moorings regularly discharged petroleum into shore waters. Even the burning of one gallon of gas by a small two-stroke motorboat engine added 105 grams of oil, 57 grams of gasoline and .6 grams of phenols to the water. Before leaded gasoline was prohibited, a half gram of lead was deposited in the water for every gallon of gas consumed.[19]

Inadequate septic systems also plagued the Cape's wetlands. The desirability of homes with waterfront or water-view property increased housing density on land closest to the water. Much of this development occurred before the passage of legislation protecting wetlands. As early as 1957, a study by the Harvard University School of Public Health found that the average Cape household needed at least forty thousand square feet, or approximately one acre, for on-site septic systems in order to prevent nitrogen overload.[20] The 1978 Water Quality Management Plan also recommended one acre as the minimum lot size for a septic system in order to prevent nitrogen overload of wetlands. The Environmental Defense Fund countered that the evidence from the Water Quality Management Plan itself suggested at least two acres per household.[21]

Shoreline density was far greater than one household per acre, let alone the two that the scientific reports suggested. Wastes from these septic systems, which were by their very nature too close to the water table to prevent the migration of nitrates into the wetlands, increasingly compromised the Cape's harbor, bay, and ocean waters. As shoreline property became more expensive, developers turned their attention to the Cape's many kettle ponds. Summer homes began to crowd around these delicate pools of water. Cape ponds, particularly those on the Otis Air Force Base and in the towns of Falmouth, Mashpee, Bourne, Chatham, Sandwich, Barnstable, and Bourne, began to suffer from low levels (that is, below six ppm) of dissolved oxygen. Signs of accelerated eutrophication, such as algae blooms, foul-smelling algae die-offs, fish kills, and over-abundant plant growth, began to show up in the ponds. Such unwelcome events demanded that attention be paid to the Cape's water system. James McCann, in a 1969 study of Cape ponds, warned that although the ponds and lakes were "basically clear," eutrophication of some of ponds was "being artificially accelerated due to pollution."[22]

Unlike the saltwater shoreline washed by tides, the Cape's ponds are closed systems. Rain and some water plants supply oxygen to the ponds, but decaying matter takes up oxygen. Even the healthiest ponds experience eutrophication as more and more organic matter overwhelms the available oxygen and builds up on the pond bed. Additional organic material flowing into the ponds accelerates this process.

The threat to the Cape's fresh water was increasingly visible in its ponds. Visitors to Centerville's Long Pond in 1963 would have found a sandy beach, clear water, and a healthy plant-to-fish ratio. Returning to that pond ten years later, visitors would not have a sandy beach on which to sit, nor would they want to swim or fish in the water. The pond water was cloudy, and up to four feet of debris and muck lay on the bottom. The fish and healthy plant life were all but gone.[23] A 1976 study found that although the levels of coliform bacteria (a type found in intestines) jumped during the summer in most Cape ponds due to bathing and overloaded shoreline septic systems, the levels were for the most part not high enough to discontinue recreational activity. On the other hand, the researchers pointed out that the ponds were demonstrating significant evidence of increased eutrophic conditions. A 1979 study by the Association for the Preservation of Cape Cod (APCC) found increased nitrate and phosphate in groundwater; increased suspended solids in ponds and streams; excessive plant growth and oxygen deficiency in ponds; and bacterial contamination everywhere, along with heavy-metal salts such as copper, mercury, arsenic, and lead. Although the APCC found that most of these problems were at an early stage, it nonetheless urged that steps be taken to ameliorate these conditions.[24]

Algae blooms and elevated chlorophyll levels, associated with increased nutrient loading from septic systems and lawn and garden fertilizer, became more common in the last quarter of the twentieth century.[25] The Mashpee River had thirty times the allowable nitrogen levels.[26] Leachate from domestic sewage effluent grew, while runoff from parking lots and roads added oxygen-demanding wastes into the ponds and provided the nutrients for excessive plant and algae growth. As this plant life died and decomposed, it took up dissolved oxygen, further depleting the supply of oxygen in the water. More Cape ponds registered low levels of dissolved oxygen, less than six parts per million. Several ponds, particularly those located near higher-density populations, registered dangerous levels of fecal coliform.[27] While low levels of dissolved oxygen accelerated eutrophication, high levels of fecal coliform indicated a health risk to bathers.

Although ponds may have been threatened, they were not the only Cape waters in trouble. By 1976 the Water Quality Management Plan found that Barnstable County was "in an early stage of groundwater degradation."[28] In December 1978, the *Cape Cod Times* reported that the Wellfleet Board of Health was notified that it had six months to clean up sources of pollution, mostly from inadequate septic systems, in and around the harbor at Duck Creek. If the town failed to meet this deadline, its shellfish areas would be reclassified as "prohibited" and shut down for five years. The town responded by cutting the flow of water from Mayo Creek to the harbor and requiring septic system upgrades for the worst offenders. Although these actions abated the problem enough to placate state officials, the problem of pollution in the harbor continued to haunt the town, especially given the growing importance of its oyster crop.[29]

On top of increased nitrates from septic systems, excessive fertilizers used on lawns and golf courses also overloaded the wetlands. And they were not the only threats to Cape waters. Pesticides and other toxic chemicals had increasingly become the weapons of choice for homeowners who were unaware of alternatives in their battle against insects, mice, mold, and other living entities that tend to thrive in the damp Cape climate. These chemicals also compromised the Cape's mostly underground freshwater system.

CAPE COD'S SINGLE AQUIFER

The story of what was happening below the surface was ominous. The Cape's fresh water, in its ponds and wells, was the canary in the mine for those willing to listen. Although local concern was raised in the prewar years about potential problems for the Cape's supply of fresh water, not until the postwar period did this concern begin to focus on the whole of the Cape. Underlying the Cape is one large aquifer, as the EPA designated it in 1982.[30] Its groundwater and water table are the same water system. Fresh water enters this system through rain. Rainwater seeps down into the water table, keeping saltwater infusion at bay. Because salt water is heavier than fresh, the Cape's freshwater "lens" sits on top of brackish water. This freshwater lens thins and approaches the surface as one nears the water's edge. Freshwater recharge pushes the brackish water down and outward. Although the Cape averages forty-three inches of rain and snow yearly, after runoff and evaporation, only an average of sixteen inches enters the recharge area. In an average year Cape residents and businesses can draw down that amount of water from the water table without significantly threatening saltwater infusion. When fresh water flowed out from the center of the aquifer it was balanced on the periphery of the Cape by the inward diffusion of salty water. These hydrodynamic shifts are continuous and seasonal, but with greater withdrawal of fresh water from the aquifer the inward diffusion of salty water increased.

The demand for water from the Cape's aquifer grew exponentially in the postwar years, especially in the last quarter of the century. Homeowners increased their use of washing machines, garbage disposals, and garden hoses while hotels and motels used dramatically more water to maintain lawns and swimming pools and to support visitors' customs of long showers. The extensive watering of golf courses also pulled more and more water from the aquifer. By the end of the century investigators for the Water Quality Management Plan estimated daily per-person water use at over one hundred gallons, more than double the 1960 average.[31]

Water used and then flushed into septic systems or cesspools, although it may have generated an overload of pollution, did reenter the water system. Likewise, when runoff from paving for roads, parking lots, and expansive roofs was directed into the ground through water gardens, French drains (trenches filled with loose stones and covered with earth), or freshwater marshes, the water, often polluted, generally returned to the aquifer. Storm sewer systems and wastewater sewer systems with ocean outflows, on the other hand, led to a drop in replacement water returning to the system.[32] The practice of sprinkling and watering lawns, gardens, and golf courses during daylight hours led to a loss of water to the system through evaporation. The result was a drawdown from the underground water supply and an increase in saltwater infusion pushing into the aquifer. Those taking water close to the meeting point of fresh and salt water drew water with increasingly higher salt levels. The reduced water entering the watershed also lowered the level of the water table, shrinking the size of the kettle ponds.[33]

Because a single aquifer underlies the Cape, problems with its quality created by collective use and abuse affected everyone. The pollution, once it entered the aquifer, migrated slowly away from its source point.[34] Each individual family's or business's contribution to the problems seemed small, and few people wanted to take on the costs of remedying problems that seemed distant to their individual contribution.[35] All the towns and all the Cape inhabitants were involved, but not all in the same fashion.

By the end of the century, Cape towns (except for some on the Outer Cape), developed municipal water systems, although the town systems did not serve all residents. For those without municipal water, private wells on the homeowner's property provided their water. Most Cape towns had a combination of each system with town water supplied to the more densely settled areas and private well water supplying the more outlying, less densely settled areas.

PUBLIC WATER VERSUS PRIVATE WELLS

Much of the Cape was already plotted out into lots smaller than either the one-acre suggestion of the Water Quality Management Plan or the two acres scientific studies suggested in order to safely dispose of wastes through individual septic systems. Towns therefore looked toward other means to alleviate the nitrogen overload on wetlands and the underlying aquifer—a solution that would also avoid resistance from builders, who opposed a change in zoning to accommodate the increased lot size. Towns that were under the greatest pressure from development (particularly on the Mid-Cape) responded by buying hundreds of acres of open space to protect water supplies. But even with these open spaces, problems of water quality continued to haunt the Cape.

A 1976 study of Barnstable County wells found limited indirect recycling of water discharged from on-lot septic waste systems, but noted that the "continued increase in density of septic units . . . will lead to pronounced pollution of shallow groundwater, particularly in . . . areas with lots of ten thousand square feet or less."[36] A 1978 study found that generally the Cape's groundwater tested above drinking standards, but there were several localized areas of degradation due to onsite pollution from septic systems and salts.[37] Saltwater infusion and organic pollution in some coastal areas forced towns to give up private wells in favor of public water. Beginning in the more densely settled areas on the south coast, towns laid out protected well fields and began drilling wells and building pumping stations and water towers. By 1990 the heavily populated towns on the Upper Cape, and Provincetown on the Lower Cape, had municipal water, whereas those towns with lower population density had either a combination of municipal and private well water or primarily private wells. Wellfleet's town center and harbor area confronted significant saltwater infusion, which the town dealt with initially by having commercial establishments import water. Finally in the new century the town built a municipal water system for its center.

In 2011 the Cape had a total of 125,800 single-family residents. Public water systems served 85 percent of them. The rest was served by private wells.[38] These municipal systems relieved the problems of organic pollution and saltwater infusion in individual wells, but were themselves dependent upon the Cape's single aquifer. By the end of the century, increased population density threatened even protected well fields. This situation forced Provincetown to look for its water in the neighboring town of Truro and ultimately on the lands of the Cape Cod National Seashore, when its local wells registered over fifty milligrams of salt per liter of water, well over the twenty milligrams per liter allowed.[39] In order to protect their water supply, some towns, such as Barnstable, Harwich, and Chatham, began building a municipal sewage treatment plants.[40] Yet into the late 1970s more than 90 percent of the Cape's year-round population relied on individual cesspools or septic systems.[41] By 2000, 85 percent of Cape homes still relied on traditional individual systems.[42]

The dramatic increase in housing, as well as the large numbers of visitors across the Cape, quickly pushed existing methods of disposing of non-trash waste, particularly human and chemical waste, beyond existing capacity. Average production of nitrogen was seven pounds per person per year. In the best situations a three-person dwelling would need two to three acres per dwelling, a significantly higher requirement than the one-acre suggestion of the Harvard Study or the Water Quality Management Plan. Many of the homes built between 1935 and 1955 were summer cottages, and their septic systems, even in the best of times, could accommodate only seasonal use. Empty lots and open space were scattered among these cottages. Although during the summer months a septic system or cesspool might temporarily overload, the long off-season and empty space could absorb much of that overload. The dramatic increase of development in the second half of the century and the conversion of seasonal dwellings to year-round dwellings contributed to septic system or cesspool failure across much of the Cape.[43] The Water Quality Management Plan found that the area along Lower County Road on the south shore suffered from significant pollution and sewage seepage that had been made worse by the conversion of summer homes to year-round homes. The WQMP recommended that Dennis create a "seasonal residential district" to prevent future conversions of seasonal dwellings. The WQMP also recommended such a plan for Eastham's western shore because its densely developed seasonal dwellings using cesspools were already overtaxing private on-site systems.[44] Along the coast, and in many of the new developments, often lots of ten to twenty thousand square feet or less were typical. Because the water table is within ten feet of the surface near the shore, septic systems discharging their wastes in that shallow range threaten the surrounding water.

SOLID WASTE DISPOSAL

The regrowth of the Cape's forest led to the appearance of environmental healing, but that healing could not hide the growing mountains of waste that accumulated in the prosperous decades following World War II. Increased population density increased tax revenues, but also strained the towns' ability to dispose of their wastes. For most of the nineteenth and early twentieth century Cape Codders either recycled or burnt their wastes. Bottles were returned to stores, and household trash and scrap lumber as well as brush and garden waste was regularly taken to a fire pit or old barrel and burned. Food scraps were fed to chickens or pigs. What could not be burned or fed to animals was buried. Over the course of the year, very little waste needed to be taken to the town dump, which emerged in the early twentieth century commonly at an abandoned sand pit. In the postwar period fewer Cape residents had animals to feed their kitchen scraps to, or the room to bury wastes, or neighbors willing to accept regular burning of trash. Visitors had little opportunity to burn or recycle their garbage. Increasingly, those who lived on and those who visited Cape Cod found it more difficult to recycle their wastes and had a vastly greater amount of waste to dispose of. Construction by professional builders, whose types of material multiplied, generated more construction waste; most of it ended up in town dumps. The growth of chain stores and department stores on the Cape increased the items available to purchase that came heavily packaged, adding more material to the growing waste stream. Although the Falmouth dump was already facing overflow by midcentury, until the late 1960s most towns' existing dumps adequately accommodated the increase in waste volume. But as the population of both visitors and residents grew, and the per-person waste generated increased as well, traditional methods of dealing with waste on the Cape began to break down. Waste disposal is necessary for the continued flow of commerce. Consumed goods need to be disposed of to make room for new goods. But there is no ultimate sink, as Joel Tarr has reminded us.[45] And on the Cape the sink was small and shallow.

A 1978 Association to Preserve Cape Cod (APCC) report noted that by the

1970s all the towns on the Cape faced problems with solid waste disposal. Landfills used by Cape towns consumed between fifty to two hundred acres of open land and were reaching their capacity. None of them were capable of meeting the state requirements for landfill operations. In 1975 the Federal government passed the Resource Conservation and Recovery Act to control pollution generating landfills, adding further pressure on Cape towns to find an alternative to local dumps. New landfills were required to have impermeable underlining to prevent contamination leaching into groundwater. This added to the expense and problems of maintenance of the landfill, but, even with the new requirements, landfills sent out a plume of contamination. Some towns, such as Wellfleet, operated incinerators to cope with their waste problems, but air and waste pollution problems with this method led to its abandonment in the early 1980s.[46] By 1990 towns across the Cape transformed local dumps into transfer stations and moved waste off-Cape for disposal.[47] Changing the name from dump to transfer station did not change the consequences of past usage. Old dumps continued to leach pollutants into the aquifer. In Eastham wells of homes near the old dump registered levels of 1,4-dioxane—a colorless liquid compound, probably carcinogenic to humans, and primarily used as a stabilizer in solvents—which forced the town to supply the residents with bottled water.

GROWING ENVIRONMENTAL AWARENESS

Ezra George Perry and Lorenzo Dow Baker were proponents in the late nineteenth century for the shift of the Cape's economy from one of production and extraction to tourism and recreation. They grew to maturity as the regime of production and extraction declined. They finished their lives at the beginning of the Cape's new regime. What lay before Perry and Baker were only visions of prosperity and recovery. James Nickerson Jr., like Perry and Baker, was the son of an old Cape family. His father had been both an early twentieth-century developer and a supporter of conservation and restraint. The junior Nickerson was just a teenager in the declining years of the Cape's regime of production and extraction. At eighteen he signed on as a first mate on one of the last working sailing schooners. But the age of sailing ships was over, and Nickerson took up teaching and earned a law degree. He practiced law on the Cape and was elected to represent the district in the state House of Representatives. With the growing importance of tourism, Nickerson's law practice increasingly involved issues associated with development and land transfers. In his retirement, Nickerson directed his attention to preserving the Cape from the detrimental consequences of the new economy.

James Nickerson lived long enough to come to realize the new economy had its esthetic and environmental costs. He dedicated the last years of his life to protecting what he noted was "only a narrow strip, extending into the ocean."[48] Although Nickerson failed to produce a consensus around an approach to the problems he saw with the Cape's development, his belief that the Cape was indeed a fragile ecosystem, and that its role as a tourist destination would bring environmental costs, did become more widely accepted by the end of the twentieth century. This was especially so since it became a truth that was not limited to Cape Cod, but extended to many areas on the coasts, as people from other shorelines also came to the realization that environmental degradation at the edge of the water was no longer tolerable.

Concern about water quality and recreational use of coastal areas of Long Island Sound, the Jersey Shore, and the Great Lakes increased dramatically over the decade of the 1960s. In the late 1960s water quality along the western shore of Long Island Sound was so degraded that locals shunned eating its shellfish and swimming in its waters.[49] Off the shore of Staten Island a huge mass of congealed sewage known as the Arthur Kill Blob not only made bathing untenable but also discouraged residents from walking the beach. Lake Erie had long been on the receiving end of untreated sewage from Detroit and Buffalo. In the 1950s and 1960s phosphates and fertilizer runoff added to this sewage waste load, leading to huge blooms of algae that died and drained oxygen from the water. Islands of dead matter, sewage, and garbage began floating about and foamy soapsuds clung to the shore. Citizens along the lake organized a Save Lake Erie campaign. Along Lake Michigan's southeastern coast, residents formed a Save the Dunes Council to stop the building of steel mills along the shore.[50] On the California coast, the Santa Barbara oil spill of 1969 sent close to one hundred thousand barrels of crude oil into the Channel and on to the beaches. Public outrage spurred political action to control offshore drilling. Neighborhood and local organizations took shape in numerous shorefront communities to push for protecting, if not improving, water quality and coastal lands.

Prodded by sport and commercial fishermen and other groups also concerned about declining fish stocks, scientists began publicizing the importance of coastal estuaries and marshlands for fish populations. At the same time, wetland degradation and loss seemed unstoppable. Local development interests saw these areas as cheap unwanted land. Civic leaders hoped the development of this land would lead to added revenue and new jobs. Many locals believed the dredging of wetlands for marinas or development meant a reduction of bugs and smells. Realizing the power of local development interests, groups interested in protecting wetlands and shorefronts, conservationists, fishermen, birders, and scientists turned to Washington for legislation. In Washington, those concerned over wetlands and coastal areas joined others pushing for broad environmental action. Out of the larger concern came the Environmental Protection Agency (EPA) in 1970, and in 1972 Washington responded to specific shorefront concerns with the Coastal Zone Management Act (CZMA).[51]

The CZMA instructed each state with a coastline to draw up a Coastal Zone Management Program. These plans were to help the states gain control over coastal environments before they became hopelessly degraded by overuse and underprotection. The CZMA placed both the federal government and the states in direct involvement with the affairs of coastal communities. As was the case with the Cape Cod National Seashore, that involvement drew different reactions. For some, it was a necessary step to protect a fragile and much loved space where local bodies were too weak, conflicted, or compromised to act, whereas for others it represented an intrusion on individual freedoms and local autonomy. It certainly added to an increasingly politicized world. Following the directives of the CZMA, Massachusetts issued its "Program and Final Environmental Impact Statement" in 1978. It was not surprising that Cape Cod occupied a central portion of that report.[52]

As the report noted, "The natural resources of the Massachusetts coastal zone are among the Commonwealth's most important economic resources. Businessmen, tourists and residents alike are attracted to the coast. . . . These areas experience sustained pressures for development. . . . However the very resources which attract so many interests to the coastal zone and support myriad activities and uses are endangered. Quite often differing activities demand the same resources, the same scarce or fragile piece of land and water. But since the coastal zone is a finite resource, it is impossible to meet the needs of all the conflicting demands for uses and activities along the coast."[53]

WATER POLLUTION POLITICS

The politics of pollution abatement faced many obstacles. Concerned that zoning restrictions limiting construction lots to one acre or larger would reduce employment in the building trades, hurt the real estate industry, and limit future tax revenues, towns' zoning laws lagged far behind the recommendations of those studying water quality. Cape towns worked, not always successfully, to buy open land as a pollution buffer and as a way to protect water supplies. The proposal for "seasonal residential districts" failed to come into play. Wellfleet, for example, adopted its first true zoning bylaws in 1966 and specified the minimum lot size of fifteen thousand square feet, which was thirty-five thousand square feet short of the 1957 Harvard Study recommendation.[54] In 1978, when faced with a growing concern for water quality, the town of Barnstable, with the support of the planning board, the conservation commission, the water department's superintendent, and a majority of the selectmen, proposed to increase its zoning requirement for residential construction to two acres. Resistance from builders, developers, and real estate agents at the town meeting prevented the measure from gaining the two-thirds vote it needed to pass.[55] Yarmouth, which showed nitrate levels of two to three parts per million (ppm) and sodium salts in excess of allowable limits in two of its town wells, attempted to deal with the problem by proposing to set aside twenty-eight acres of land for protection. Even this watered-down proposal—far less than the 5,325 acres suggested by the Water Quality Management Plan—went down to defeat, failing to get the necessary two-thirds majority in the face of opposition from the development interests.[56]

Some Cape towns began using their zoning powers to establish water protection districts around public wells, but more was needed. Without more aggressive action, water pollution continued to plague the Cape. Town wells had to be abandoned because of pollution. Homeowners in Falmouth were advised to drink only bottled water because of polluted well water. Beaches were closed due to sewage.

It was not only human waste that was threatening the Cape's fresh water. Most of the newer large and expensive homes contained all of the conveniences of modern life: washing machines, dryers, dishwashers, air conditioners, garbage disposals, security systems, and even Jacuzzis. Their surrounding grounds reflected luxury as well. Fescue lawns replaced poverty grass, and expensive bushes and ornamental flowers replaced poison ivy, honeysuckle, heather, and

Rosa rugosa. Lawn services came regularly to water and tend the lawns and gardens, pouring on heavy doses of fertilizers and chemical herbicides and pesticides. Initially, DDT until it was outlawed in 1972 and then diazinon, a common household pesticide, were used to control garden and home insects. Recommended through advertising as a means of controlling ticks and Japanese beetles, chemicals were freely applied to lawns and shrubs despite growing concerns about the toxic nature of these treatments.[57]

In addition, hoping to present a more neighborly suburban appearance, shopping centers, strip malls, motels and hotels, and restaurants established plantings around their commercial operations. In a region with high labor costs during the growing season, owners of these establishments were quick to substitute chemicals for labor as a means of pest and weed control. And although golf courses on the Cape originally appealed to players because the Cape landscape resembled that of coastal courses in England, Wales, and Scotland, by the mid-twentieth century golfers associated a good-looking course with smooth green grass fairways. The Cape's environment did not favor such an appearance. Aided by strong fertilizer additives and heavy watering, it could.

By the 1970s fescue lawns supplanted the Cape's natural flora: wild beach grasses, poverty grass, bunch grasses; heather, blueberry, huckleberry, and bearberry bushes; and wild roses. In 1972 Monica Dickens, a long time Cape resident, noted traditional Cape Cod grasses were not very green but that emerald green lawns were flourishing, thanks to chemical products whose "runoff has added to the degradation of many an inlet and marsh."[58] The 1978 Water Quality Management Plan noted that in addition to golf courses, fertilizer for lawns and gardens was a "significant source of pollution."[59] As with water table management, fertilizer and pesticide pollution rested within a confusing politics of "everybody's problem and therefore nobody's problem."

Chemical additives to lawns polluted Cape's waters, wetlands, and marshes, but they were not alone in so doing. Cleaning services and homeowners liberally applied oven cleaners, bug sprays, furniture polish, bathroom disinfectants, bleach, cleaning solvents, and drain cleaners to achieve a home that met the standards of modern living. The household drain commonly received these materials while motor oil, antifreeze, paint, and paint solvents continued to be dumped in out-of-the-way ditches or town landfills until the late 1980s. From household drains and landfills these chemicals leached into groundwater. Although the Cape's economy shifted away from agriculture and production, as rising expectations of cleanliness increased, household use of chemicals meant more toxic pressure on the Cape's water system.[60]

THE POLITICS OF WASTE

The 1970s Water Quality Management Plan recommended public sewage treatment systems for nine towns and locations: Buzzards Bay/Wareham, South Sagamore/Sandwich, Yarmouth, Chatham, Barnstable, and Provincetown/North Truro. Barnstable and Chatham initiated systems, while the others resisted. The Water Quality Management Plan of 1978 called for the Bass River section of Yarmouth to draw up plans for and to build a sewage treatment facility in five years. The town refused. In 1978 the commissioner of the Department of Environmental Quality Engineering (DEQE) informed Falmouth that it needed to implement a town sewer system. The DEQE notified Falmouth of its violation of Title 5 of the Massachusetts State Environmental Act, noting that it was discharging raw sewage at Woods Hole.[61] Informed that a sewer system to abate these problems would cost each resident four thousand dollars over ten years, residents objected to taking action. The town meeting voted for a minimum approach, short of compliance with the DEQE order. The DEQE backed off from taking further action against the town, even though it was not in compliance, because it had at least taken some abatement action.[62]

Cesspool and septic system failure were not the only problems associated with private on-site systems. Most of these systems needed periodic pumping out to prevent their backing up into homes and businesses. The sewage, once removed from the site, needed to be disposed of somewhere. Typically it was taken to town dumps and poured into open pits—which were illegal under both state and federal laws by the 1970s but still widely used—or put in lagoons, which were approved on an interim basis. Although Barnstable and Chatham accepted the wastes at their treatment plants, most waste on the Cape did not undergo any formal treatment. Seepage into the water systems added an additional pollution burden to the region's water system.

Dependence upon a single aquifer for its water supply raised other concerns. It also raised the necessity of trans-town cooperation. Wastes from cesspools, septic tanks, road runoff, municipal dumps, and household fuel tanks enter the groundwater through gravity percolation. Once there, they slowly move into the surrounding water table. As water is drawn out of the water table, the wells draw polluted leachate, including salts, toward the well. Although the Cape gets enough rainfall to recharge its water table, increased use of water on the Cape draws down the water table, pulling pollutants away from their source into the larger aquifer upon which Cape Cod depends for its drinking water.

Saltwater Infusion

Although concern over polluted wells was raised in the 1970s, town wells were initially located on land far from major sources of pollution and private wells were seen as private affairs. Saltwater infusion was different. Given that the Cape's aquifer is surrounded by salt water and that saltwater infusion was more noticeable and widespread, salt was the first pollution to set off alarms.

Infusion was not the only salt threat to the groundwater. Charged with keeping the roads clear of snow and ice, state and local DPWs mobilized at the earliest threat of snow or ice to dump salts on the roadways. With its integration into the larger New England economy, the Cape was heavily dependent on its arterial roads, particularly the Mid-Cape Highway and Routes 6A, 132, and 28. Cape Codders no longer produced their own goods. Food, clothing, fuel, and emergency supplies came from off-Cape, and they came by truck. With few major highways supplying the Cape with the goods it needed to survive, local and state DPWs took responsibility to have these roads drivable and as accident-free as possible. From their perspective that meant dumping enough salt to keep the roads clear of snow and ice. To do so, they maintained on the Cape huge mounds of salt, much of it only partially covered, or totally uncovered.[63] Runoff from these mounds of salt eventually ended in wetlands or groundwater. Once the snow started to fall, so did the salt. The DPW's formula for salting roads was three hundred pounds of salt for each mile per lane of highway. A mile of the Mid-Cape highway received twelve hundred pounds of salt with each application, and a severe storm led to many applications.

Although towns attempted to deal with the problems of pollution and salt with municipal water systems, salts in the groundwater threatened even these systems. A 1981 study found that several public town wells were approaching the twenty milligrams per liter limit for salt while many private wells were well over the limit.[64] The increased salt in local water supplies led local towns to instruct their DPWs to mix salt with sand and to lobby the state DPW to reduce its salting. Steve Roy of the state division of water began pushing the

DPW to reduce salting in 1978, but the DPW refused to budge, claiming that its responsibility to keep the roads clear and limit accidents trumped environmental concerns. After an agreement between the Department of Environmental Quality Engineering and the DPW to experiment with less salt was scuttled by the DPW, the legislature attempted to force DPW to reduce salt by giving the DEQE more power to reduce salt in sensitive areas. Despite the intervention of the state legislature, the DPW ignored or sidestepped the DEQE's attempt to limit salting and continued with only limited moderation.[65]

Military Wastes

A toxic waste plume flowing out of Camp Edwards and the Otis Air Force Base (now called the Massachusetts Military Reservation) on the Sandwich-Bourne-Falmouth border also threatened the Cape's water system. Otis Air Force Base covered over thirty square miles of mostly undeveloped land, which is the water catchment area for much of the Upper Cape. Being within easy commuting distance of most of the western Cape, Otis provided a significant boost to the area economy, providing 20 percent of the Cape's year-round employment by the late 1960s. But Otis had a negative impact on the Cape's ecosystem.

Toxic wastes—from aviation oil and fuel, defrosting fluids, chemicals for cleaning and servicing aircraft, leaking holding tanks, and runoff—leached into the aquifer and began to spread outward from the base.[66] By the 1970s several of the wells on the base were shut down because of phenol or nitrate contamination. Nitrate pollution from the base's septic treatment plant turned up in Falmouth.[67] By 1980, with the uproar over Love Canal in 1978 spurring national awareness of the health impact of industrially polluted water, attention focused on Otis and it was added to the Super Fund list as the Massachusetts Military Reservation Super Fund Site.[68] Since the mid-1980s, attempts at remediation have been made at Otis under the supervision of the U.S. Environmental Protection Agency, but with little success, as the toxic plume continues to spread outward from the base.[69] Investigators found seven separate plumes linked to the site. The plume flowing out of the sewage treatment plant moved southward at a rate of one foot a day, reaching Falmouth's residential and commercial center in 2000. To the north, another plume is flowing into Sandwich and has already contaminated a well in that town. Although a link between the increased cancer rates in Barnstable County and the toxic wastes flowing out of

Otis was not proven, concern over cancer did increase public awareness of toxic wastes entering the aquifer.

Toxic waste also flowed into the aquifer from gas station underground petroleum holding tanks, which when leaked into the surrounding soil led to the closing of town and private wells. At the other end of Cape back in 1977, leaking petroleum from a Provincetown gas station compromised 60 percent of the town's water supply.

LOST LAND

Pollution and environmental abuse have gained significant amounts of media coverage over the last thirty years, but the Cape has been struggling with another dramatic environmental issue for far longer. Each year nature reclaims more of the Cape. Parking lots are abandoned, and beachfront homes have washed into the sea. In 1996 the Highland Lighthouse in Truro and the Eastham Nauset Lighthouse had to be moved back from the cliff's edge. The lighthouse in Truro was 330 feet from the edge of the bank when Thoreau visited in the 1850s, yet he foresaw that the erosion of the land might force its eventual movement. These are not the problems of erosion caused by overgrazing or cutting down trees; these are the problems of the ocean itself.[70] Humans may play a role in this: building roads and driving and walking across dunes destroy beach grasses, making wind and water erosion more severe. But even without a single road or beach path, the Cape would still lose its shoreline. Global warming increases the intensity of ocean storms and raises water levels, but Cape Codders themselves play but a small role in this. Their major human role centers on where they build their homes and what they do to protect them.

Beach erosion was not new to the twentieth century. Wind and wave actions continuously take and redeposit sand. Sandbars across the Cape have been in continuous motion since the days of the last glaciers. Channels and harbors have opened and closed; and islands, such as Billingsgate off Wellfleet, have come and gone. Seven thousand years ago Georges Bank was a giant island protecting the Cape from eastern storms.[71] Rising sea levels submerged the bank, exposing the Cape's eastern front to the full force of an angry Atlantic Ocean, moving sand both north (creating Provincetown) and south (creating Nauset and Monomoy). More recently, ocean-side cliffs have lost on average two-and-a-half feet a year—lately, three feet—while the bay side lost an average of a foot each year.

Unlike the lighthouse keeper to whom Thoreau talked over 150 years ago, those with property near or on the shoreline in the twentieth century did not expect to move their homes. Sand moves. Human development on these shifting sands represented invested capital and tax revenues. When storms and normal wave action took beachfront property, owners demanded action. Between 1945 and 1970 over ten million dollars was spent to try to limit beach erosion on the Outer Cape. In 1966 the Cape Cod National Seashore (CCNS) dumped three thousand cubic feet of rubble along the bank and beach at Coast Guard Beach to protect the road and parking lot associated with the beach. The efforts gained the road and parking lots a few years of life but at the cost of what the superintendent of the CCNS called a "hazardous and unsightly situation."[72] Following the failure of this 1966 attempt, the National Seashore adopted a "live with erosion" strategy.[73] In fact, the CCNS lost a bathhouse and a parking lot at Coast Guard Beach in 1978, thus limiting its capital infrastructure on the coast and making a "live with erosion" strategy affordable. But the buildup of motels and expensive homes on land outside the CCNS represented heavy private investment. Threats to that investment brought town and individual action. Faced with bayside erosion in 1973, Provincetown allocated funds for building even more groins (structures of stone or cement stretching out from the shoreline into the water) at Beach Point; the groins slowed the motion of sand by holding the sand flowing toward them, but on the other side of the groins sand eroded. Homeowners and/or towns have also tried building barriers of stone, cement, and wood to limit the erosion and protect shorefront property. These solutions proved to be expensive and were only successful in limited cases where the erosion was minimal. Towns have also dredged sand and pumped it onto the shoreline to expand the beach. Although this was a short-term solution, it was less expensive than building stone or pillar groins and walls. The dredging and pumping of the sand itself nonetheless caused significant disruption of delicate marine plant and animal ecosystems.

Despite concerns about water quality, congestion, and development, the Cape's surface still retained much of the beauty Henry David Thoreau noted so many years ago. That beauty continued to draw thousands to the Cape's beaches and ponds. Although many Cape Codders bemoaned the proliferation of contrived Cape-themed restaurants, mini-golf courses, and outlet stores, for many others those things made up a good vacation and were part of the Cape's appeal. As the Cape's tourist economy prospered, the costs of being on the Cape also rose, not only for those buying homes, but also for those visiting. The high price of land and taxes across the whole of the Cape drove up the price of rooms for those staying there. Visitors to these expensive rooms expected more than the simple accommodations the older motels offered. Guests expected central air conditioning, heated pools, game rooms, large color televisions, green and well-manicured lawns and gardens, lounges, conference rooms, and restaurants with all the services and offerings one would expect in an expensive hotel. These new motels needed far more labor than could be supplied by locals. Seasonal international students increasingly replaced local students as cleaning and maintenance staff. High land values and taxes, as well as the increased value in capital stock that these motels represented, pushed them to stretch out their costs over more than the two summer months of the high season, making labor, as students returned to classes, all the more difficult to find. By the end of the century these larger establishments looked to a more permanent but lower-cost labor force, which often meant immigrant workers from Haiti, Africa, and Latin America. Many of these new Cape workers found housing in decaying old motels or in the more run-down sections of Hyannis or Falmouth, giving both areas a cosmopolitan flavor they had not experienced since the days of the multinational whaling crews.

A new politics of environmentalism found major expression in the Cape Cod Commission, but whether public concerns combined with local, state, and federal politics can adequately protect the environment of the Cape remains for the future to tell. James Nickerson's father, James Nickerson Sr., of the turn-of-the-century Cape Cod Improvement Association, would have understood the sentiment behind the Cape Cod Commission. But by the time the Coastal Zone Management Group issued its "Program and Impact Statement" and the CCC issued its "Prospect: Cape Cod," the Cape had already eaten the economic fruit of tourism. It was unclear if the lost Eden of unspoiled beaches, open vistas, secluded places, picturesque and quaint villages, and clear and clean ponds, streams, and harbors could be recovered or preserved. Although public opinion polls and votes on issues such as the creation of the Cape Cod Commission indicate that a majority of Cape Codders today want action to protect both the Cape's environment and its classic ambiance, in the real world of varied visions and politics with conflicting interests and resources, it is unclear what the future will bring.

CONCLUSION

The collapse of the Cape's second regime—the one based on extraction and production—took place over the long fifty years at the end of the nineteenth century. No single event triggered the end of that regime, and elements of this older regime continued on into the new regime of tourism and recreation.

THE EFFECTS OF EXTRACTION AND PRODUCTION

The Cape's economy, after all, had not been built upon a single industry but upon clusters of interrelated activities. Fishing, to be sure, was key, but it was supported by salt making and boatbuilding, each of which were linked to a variety of other undertakings such as barrel making, sail making, and rope making, as well as warehousing and shipping. Even the fishing industry was varied. Cod was of course the keystone fish for New England fisheries, but whaling, mackerel fishing, catching herring for food and bait, and shellfish harvesting were also important, and the relative importance of one sector of the fishing industry versus another shifted over time.

Not all Cape Codders fished. Farming played an important part of the local economy throughout the second regime. As with fishing, some crops were more important than others, and that also shifted over time. At the beginning of the nineteenth century, Cape farmers planted wheat, rye, oats, and vegetables, maintained orchards, and raised chickens, pigs, and cattle. They consumed most of what they raised but shipped off-Cape surplus grain, beef, and pork, as well as their harvest from the sea.[1] Although Cape Codders continued

CONCLUSION 199

to grow food for local consumption, their production of cereal crops declined as cheaper grain from upstate New York and the emerging Midwest pushed into eastern markets. Cape farmers, like many other New Englanders, looked to sheep as a source of a marketable commodity. This proved to be a short-lived success story, since the requisite mills and shearing facilities, as well as sheep grazing on delicate beach grasses, severely damaged the environment. By the end of the nineteenth century, Cape farmers planted vegetables and cranberries both for their families and for families on the mainland. They replaced sheep and beef cattle with milk cows, chickens, and ducks for sale to local resort hotels and to off-Cape consumers. By the early twentieth century cranberries, strawberries, asparagus, and turnips kept Cape farmers busy in the fields, but there were fewer and fewer of them each year.[2]

Although the decline in fishing and agriculture had its roots in the environment, other factors came into play in the decline. These often masked the environmental factors. In 1922, when the Massachusetts Department of Labor and Industries issued *Population and Resources of Cape Cod* in response to the growing concern about the Cape's loss of population, it reported: "The railroads brought western commodities into competition with home products . . . [leading to] a great decline in agriculture. [But the decline] was also due in part to the exhaustion of Cape soils."[3] This pattern of multiple causes of decline can also be seen in salt making and boatbuilding.

It would misread history to lay all the blame for the decline of the Cape's regime of extraction and production on one single environmental factor such as the deforestation of the Cape. But one must also be mindful that in the highly competitive, fluid, and often unstable capitalist economy, small shifts in costs engendered by environmental overuse often precipitated a more general decline without being obvious to those participating in that process.

Custom, habit, time, and resources invested in traditional economic activity encouraged Cape Codders to continue as long as they could in the economy of production and extraction. These seafarers, salt makers, boat builders, and farmers were a tenacious and adaptive lot. When whales were gone they went looking for fish or bananas. Builders of schooners and fishing catboats switched to building sailing skiffs and racing sailboats for the new well-to-do summer residents. Farmers gave up cereals for vegetables, or gave up raising beef cattle in favor of milk cows, chickens, and ducks.[4]

Farmers such as Anthony Marshall's father, with access to rich meadows of sea grass and open pastureland, stayed on-Cape producing fresh milk, cream, and butter for local hotels and guesthouses. Ultimately fresh milk from the mainland pushed even these dairy farmers to look for alternative means of extracting a living from their land. Anthony Marshall's father turned his focus to truck-farm crops, especially asparagus and turnips, and he worked part-time building and plastering new homes for summer guests.[5] The coming of electricity and refrigerators and large grocery stores pushed even these adaptive farmers to look for other sources of income. Farming required hard work and a mix of skills. Many farmers, such as Anthony Marshall's father, put those skills to work servicing the new visitors to the Cape. Those fortunate enough to have farms located near a body of water turned their land into profitable building lots.

As the twentieth century wore on, the space for traditional Cape activity shrank. This was reflected in the fishing industry. This industry, so important for much of the Cape's economy over the whole of the second regime, proved to be a wobbly mainstay. Mackerel moved on- and offshore throughout the nineteenth and twentieth centuries, engendering ups and downs both for close-shore fishermen and for those who fished for mackerel further out to sea.[6] But on top of these particular shifts, it was the larger pattern of overfishing—which led to a decline in stock—that put a hole in the mackerel fishing fleets. Cod fishermen also had to make adjustments, and some of these hastened the eventual decline of the industry. Facing competition for cod, fishermen out of Provincetown's deep harbor switched from hand lining to trawling for cod in the 1930s. Although the trawlers were able to pull in significantly more fish than the old hand-liners, the dragnets were hard on the fish populations and did not discriminate in size or marketability of the fish caught. The new technology of diesel-driven fishing boats and new nets kept fishing profitable for Provincetown fishermen until foreign competitors (with even larger fishing operations despite the ever-smaller fish stocks), undercut the town's fishing industry. Partial relief came in the form of the Magnuson-Stevens Fishery and Conservation Management Act of 1976, which limited the competition from foreign fishing fleets. As part of the economic recovery of the 1980s, banks, encouraged by the government, offered low-interest loans for American fishermen to upgrade their fishing operations. It was a rebuilt, heavily capitalized, and debt-laden fishing fleet that headed out to the traditional cod fishing grounds by the end of the century. This was a fishing fleet capable of bringing fish stocks to the edge of destruction. The heavy debt and capital cost of the new fishing boats pushed the fishermen to pursue as much fish as the technology could accommodate—that is, far more than the fish stocks could tolerate.[7]

THE EFFECTS OF A SHIFT TO TOURISM

Cape Codders, fishermen and farmers alike, shifted the focus of their activity increasingly toward visitors. Fishermen who could not afford the new, heavily capitalized boats, which increasingly dominated the fishing industry, switched to scalloping or gave up commercial fishing entirely. Some took out sport fishermen in charter expeditions. Others collected oysters and clams from the new town-issued grants. For most, the end of the twentieth century represented the end to a life at sea.

The new economy provided jobs and opportunity. Clamming shacks became restaurants, and old woodlots became camping grounds. Land was sold, and cottages, then homes, and then mansions were built. Estates needed gardeners, and cottages and motels needed landscapers. Roofs needed shingling, wells needed drilling, and septic tanks needed emptying. Restaurants needed servers and dishwashers; motels and guest cottages needed cleaning services. For some, the new tourist economy offered the opportunity to start their own companies, whether building houses, cutting trees, cleaning septic tanks, or drilling wells. For young people and those without land to develop, or hotels, campgrounds, or restaurants to own or manage, the new tourist economy offered employment at wage labor.

These new businesses and those employed by them needed the Cape's tourist economy to remain vigorous. As it was structured through the twentieth century, vigor meant growth, more homes, more motels, more and larger restaurants, and most of all more visitors. When this process was just beginning to be understood, two writers about Cape Cod, William Berchen and Monica Dickens, warned in 1972 that, "Economic survival is said to depend on attracting tourists. But the more tourists, the less attractive the Cape until eventually, perhaps even the tourists won't want to come any more having killed the thing they loved."[8] With limited land and growing numbers of vacationers and summer homeowners, land prices increased dramatically. Higher land prices affected how the land was used and the value of what was put on the land. This was the economic dynamic that drove the Cape for the last fifty years of the twentieth century. It not only affected the visual landscape, the environment, and the quality of the Cape's water and air; it also affected who could come and appreciate the natural beauty that Henry David Thoreau found so appealing on the Cape, where "all aspects of this desert are beautiful."[9]

From the middle of the eighteenth century through to the middle of the twentieth, a diversity of people found a home on Cape Cod. Although native-born, old-stock Americans predominated, in its fishing and whaling centers one could find people from across the globe. In the little village of Mashpee, Native Americans tenaciously held together their community. Although some Cape Codders made goodly fortunes extracting resources from the land and sea, most were of the middling sort. As Timothy Dwight noted at the turn of the nineteenth century, comfort and competence were the more common experience for Cape residents. Nineteenth-century visitors to the Cape were also a mixed group. By the second half of the century some very wealthy people came to hunt or establish summer estates while others with far fewer resources came for a setting of comfort and tranquility. Those with more modest means stayed in simple guest homes, camped, or built small cottages. The cottages Lorenzo Dow Baker built in 1902 at Corn Hill in Truro, for rent to these guests of average financial resources, were simple two- or three-room affairs with a single cold-water tap and a privy in back.[10] By 1950, a rush of middle-class vacationers came to dominate the Cape. Not only were they crowding into guesthouses, motels, campgrounds, and rental cottages, but they were also building summer homes of their own. Those reaching retirement age built or converted summer homes into permanent residences.[11] The numbers of both groups drove the Cape economy from 1940 to 1980.[12] But their numbers also dramatically increased the pressure on the Cape's environment: its water, air, and natural landscape.

Faced with an increasingly compromised environment, Cape Codders, and those who loved the Cape, rallied to find ways to protect it and its natural environment. Some of this response involved land preservation, the largest instances being town land purchases, the Cape Cod National Seashore, Nickerson State Park, the Shawme-Crowell State Forest, and the Massachusetts Audubon Society reserve in Wellfleet. The response also involved more stringent standards on pollution in the water as well as increased lot size in zoning regulations, upgraded septic requirements, and municipal sewer systems.[13] These actions represent Cape Codders working together to protect this place for future generations. But the drive of the economic system works against them.

Over the last decade of the twentieth century, new homebuyers and builders had less and less land upon which to build.[14] By 2000, the land available for residential use could sustain only an 18 percent increase in housing units (and only 15 percent in its major watersheds).[15] Today, as more people are drawn to

CONCLUSION 203

the Cape and more land is protected, the number of acres available for use diminishes, and the price of land remaining on the market increases. Increased land values pushed up the use value of the land, and the price and (often) the size of what went on it.[16] The average new house size in Orleans rose from 1,800 square feet in 1990 to 2,400 square feet in 2012.[17]

Those with the greatest resources have the greatest opportunity to realize the value of the Cape.[18] As the town of Wellfleet reported to its citizens in 2011, housing prices "have skyrocketed but wages have not. There is a shortage of affordable rental or home ownership opportunities for people who are vital to our local economy and provide critical community services—our teachers, policemen, firefighters, town employees, fishermen, artisans, small business, and service employees."[19] Heavily accommodated motels and hotels replaced simple cottage motels, and larger/fancier restaurants replaced the simple clam shacks. Because the Cape is approaching a "built-out condition," new construction is shifting from sites on open land to sites where less-expensive houses that already exist can be torn down and replaced with larger homes.[20] Smaller cottages, such as those Lorenzo Baker built on Corn Hill, are increasingly upgraded and sold as six-figure condominiums. Galleries and high-priced boutiques replace old hardware stores and pharmacies. For those with the resources to enjoy the high-end restaurants and shops and the expensive rooms and homes, the Cape became a very desirable place to summer, disproving Thoreau's assumption that it would never be the watering hole of the fashionable set. But as prices continue to rise, opportunity to enjoy the Cape as a summer home or tourist destination becomes available to smaller and smaller numbers of people. At the same time, with the Cape Cod National Seashore, the Nickerson State Park, and the several campgrounds that still operate on the Cape, the Cape is still drawing over 4 million visitors a year, many of them with limited budgets.

Zoning legislation to restrict home building on small lots, combined with the preference of the wealthy for privacy afforded by large chunks of land, might suggest that the new luxury homes, hotels, and restaurants present a smaller footprint on the land than dozens of smaller cottages with their individual septic systems jumbled together.[21] Forced to build to tighter environmental standards, these new developments incorporate the latest in insulation and septic tank design, while the older cottages more often than not leaked energy and pollution. But because of the high level of appliances, technology, paved driveways, and expensive lawn care with pesticides, herbicides, and built-in watering systems, these large homes and luxurious accommodations represent a large per capita footprint.

THE FUTURE

Unlike Martha's Vineyard, sections of the North Shore of Massachusetts, or Newport, Rhode Island, the Cape in its new regime of tourism and recreation appealed to a variety of people: the wealthy, the middle class, and even the lower middle class.[22] Small vacation cottages littered much of its shoreline, while along the roads leading back from the beaches were rows of simple bungalows occupied by retired teachers, police officers, and mail carriers.[23] These people put down deep roots in the last half of the twentieth century, and they are not likely to quickly leave the Cape. The Cape, after all, unlike Martha's Vineyard or Nantucket, is relatively accessible from urban areas of the Northeast. It is merely two hours by bus or car from Boston, so visitors of a wide variety of social classes will, for the foreseeable future, continue to make the trek down to its sandy beaches and, for the younger set, to its thriving nightlife. But for those who want to lay down deeper roots on the Cape, either to live or to stay for a week or more, rising land values and home prices are increasingly pushing access to the Cape out of the reach of the middle class.[24] As the World War II generation of retirees dies, it is not at all clear that their children or others with similar resources will be able to hold on to this vacation idyll. Those who depend upon the Cape's tourist economy and who work hard against restrictions on its future find themselves being priced off the Cape by the very industry they so heartily support.[25] But the Cape itself, especially because of its fragile nature, is at risk.

Without significant restrictions on development, the problems already manifest in the Cape's waters will spread further and wider. Nitrogen overloading has caused more than seven thousand acres of shellfish beds to be closed.[26] Ponds facing accelerated eutrophication will die. Well water will become polluted. The seashore and wetlands will reek and shellfish will become contaminated. Open space will exist only in protected areas such as Conservation Trust land or the Cape Cod National Seashore, established in 1962. Parking lots, malls, and expansive commercial enterprises will dominate the landscape. As Cape communities attempt to come to terms with these problems by means of regulating development, they must be careful to consider the environment first but also to weigh the social consequences of their actions. Prices will eventually slow development. Already the Cape is feeling the impact of these economic pressures. After a half century of population growth, the Cape lost 2.4 percent of its population from 2000 to 2010, and young families with children led the decline.[27] But slowed development may not end the problems of pollution or ensure equitable access to the Cape's natural beauty.

Without a serious attempt to protect the Cape for the largest number of people, especially those with few resources, a protected Cape will be like a Rembrandt in the home of a private collector, well protected but not widely appreciated. Not only will such a Cape be the privilege of the few; history does not suggest that the environmental impact of those few will be any less detrimental than that of the many. Rising prices and development are pushing the Cape toward a future where most of it will be owned and controlled by those with wealth. Its labor force will comprise guest workers, either living in controlled boarding structures or clustered in isolated ghettoes of low-rent housing (a pattern already beginning to develop in Hyannis and Falmouth), or imported daily from off-Cape.

This scenario does not have to play out for the Cape's future. A well-regulated and controlled waste sewage system, careful clustering of development, mixing densities, limiting energy use and housing unit size, maintaining publicly supported housing, and utilizing green technology could allow the Cape to protect both the environment and its social diversity and make it available to more people. There are significant movements in these directions. The Cape is under pressure from the EPA to develop a regional wastewater treatment system.[28] Affordable-housing authorities work to create more housing for those with limited resources.[29] After 2008, the Community Preservation Act allowed for state funds to be available for land acquisition for preservation as well as for affordable housing.[30] The more available the Cape is, the more people will have an investment in its protection.[31]

The Cape Cod National Seashore serves as a model for what can be done. Many Cape Codders objected to the creation of the CCNS, and many others to this day complain about restrictions on park activity—such as the use of off-road vehicles, letting dogs run free, tramping around the endangered piping plover's nesting grounds, burning of beach fires, and nude sunbathing. Any large bureaucratic organization has trouble listening to local concerns and interests. But over the course of its first fifty years the CCNS has tried to listen to local citizens and to be a good neighbor. Such a relationship is not easy, but so far the CCNS does represent a possibility of both providing access and protecting the environment.[32] Although many Cape residents do not like all the actions taken by the CCNS, the National Park Service does attempt to base its decisions both on public input and the best in scientific research.[33]

The Cape cannot be saved, however, by protecting a few isolated natural settings such as the CCNS, the Nickerson State Park, or the Audubon sanctuary while at the same time allowing the market to determine land use for the rest of the Cape. Such a strategy will result in a Cape composed of private preserves for the wealthy, clusters of public access areas, pockets of isolated poverty, and miles and miles of unsightly development and paved parking lots. The natural world cannot be protected in isolation. As a conservation strategy, the Cape Cod National Seashore must be folded into a larger strategy, which includes a diversity of people living and working on the Cape.

Looking further back into our history we can also find models of access and protection in the restrictions and protections our colonial and early national leaders developed around the commons and common resources. Before these attempts at husbanding resources through collective common action ran into the buzz-saw of untrammeled nineteenth-century capitalism, Cape Codders managed, with conflict to be sure, to limit the taking of fish in order to protect future stocks. They restrained the building of dams to guarantee migration upstream for fish, and they limited the cutting of wood and the grazing of livestock on common land. They held land in common, and they protected the right of access to fishing and fowling sites. In 1650 when the lands around Provincetown Harbor were threatened by fishermen cutting trees for fuel and to repair boats, Plymouth Colony restricted the use of the region to two companies of fishermen. In 1654 the colony bought the land from a local sachem and held it as government land in order to control its use. After the Massachusetts Bay Colony absorbed Plymouth in 1691, the lands, known as the Province Lands, continued to be collectively held by the colony. Settlers, mostly fishermen, were allowed to occupy the land for fishing, but they were not allowed to own the land, even after Provincetown became a town in 1727, because the land was held by the colony as a means of controlling misuse.[34] In these acts our New England ancestors attempted to balance access with private pursuit of comfort, that is, to balance public good with private interest. They placed limits on the individual freedom to maximize one's wealth at the expense of the future.

In the early nineteenth century, particularly the 1820s and 1830s, when maximizing opportunity was celebrated as the American ideal of freedom and equality, the vision of husbanding resources and the collective restrictions of resource utilization were cast aside. On the Cape the consequences were river dams that limited the upstream fish migration, even heavier cutting of woods, more intense mining of the land, heavier grazing on the dunes, and the diminishing of the commons. The Falmouth Mill-dam War at the turn of the century, when the less wealthy fought the dam that would restrict fish migration, was a reflection of this shift from public husbandry of goods to privatization and individual pursuit of opportunity. The Falmouth milldam battle was not only about preserving fish. The milldam was built to provide power to a woolen mill that was to take the wool sheared from the growing flocks of sheep overgrazing on Cape dunes and to spin it for commercial sale.

The farmers who raised ever-more sheep, and the mill owners who carded and spun the sheep's wool into yarn, focused on commercial markets beyond the local community. They believed that by tapping more deeply into the local environment, grazing more sheep and damming more streams, they could extract profitable returns for their efforts. Over the next half century the consequences of those actions led to an environmental crisis that helped engender the Cape's economic crisis.

Today's dynamic tourist economy is also driving the Cape forward toward more profits and individual gain, but like the economy of the first half of the nineteenth century, without restraint it could likewise head toward an environmental crisis. If setting aside reserves of protected land is not a good enough solution, how can we use the lessons of the past to help us navigate the future? Preservation is important and tourism will be a major part—in fact, the largest part—of the Cape's economy for the foreseeable future, but so is creating a sustainable economy more removed from the tourist industry. Such an economic base, at least a step removed from development and tourism, could create a local counterbalance to the pressure for development.[35]

With sufficient imagination, creativity, and sensitivity it may be possible to build a coalition of conservationists, environmental regulators, and locals dependent upon a balanced local economy. Already there are roots of such an independent economic base. Shellfish harvesters have an interest in the protection of the wetlands and coastline. Although these people face the ongoing problem of excessive nitrogen polluting their beds, the development of a regional waste system and the improvement of individual septic systems should help mitigate some of this problem. The Association for the Protection of Cape Cod in 2010 estimated that there were 606 acres of shellfish grants already working on the Cape with more opening up every year. Between 1998 and 2010 aquaculture activity in Massachusetts rose 147 percent, and most of that is happening on the Cape. This activity generated 5.2 billion dollars a year.[36] The opening of the Herring River and other estuaries should increase the quality of the water that flushes over these oyster and clam beds while also expanding the areas available for more beds.[37] Restaurants, because of the wealth of tourists coming to the Cape, are more and more willing to pay premium prices for local fish and shellfish, which is one of the reasons that any strategy for the Cape's future has to involve a heavy dose of tourism. Fishermen working together and linked to Cape markets could develop a sustainable harvest of shellfish and hook-and-line-caught fish.[38] Local restaurants increasingly advertise local and organic produce on their menus. With the development of farmers' markets along with a renewed interest in small-scale sustainable agriculture, farmers are reclaiming Cape farmland. Presently the Cape has 4,269 acres of active farmland with the average farm size of fourteen acres. Many of these farms are new, existing only since the beginning of the twenty-first century.[39] Chicken cooperatives and small organic farms have already begun to spring up in towns across the Cape. These small-scale operations obviously cannot survive without tourists and summer people with disposable income. But they do offer opportunity for a livelihood independent of building and development, a sustaining rather than an expanding economy. With a state policy that provides tax advantages for giving up development options on farmland, a significant amount of Cape land could be protected from development and still be used productively and sustainably. Such activities offer the possibility for coalitions of small farmers, shellfish harvesters, and environmentalists to advocate for protecting the environment and for maintaining a diverse community that those with moderate incomes can still call home.[40]

Such a world would require Cape Codders to come together around a vision that is not tied to economic expansion. A quick perusal of a Cape bookstore gives a glimpse of the possibility for that future. Besides the traditional tourist guides that focus on restaurants, motels, nightlife, stores, and beaches, found on the shelf next to the older early twentieth-century Cape literature focused on quaint villages and homespun characters, there is a growing body of literature about the Cape's environment. Writings on birds, plants, fish runs, and seasons have become popular. More and more visitors to the Cape are seeking out nature paths, whale watches, and out-of-the-way streams and marshes.[41] These visitors require infrastructural support—parks, reserves, conservation areas, walkways, portable toilets, guides, canoes, and kayaks—but their activity leaves a smaller footprint on the ground and does not require so much growth as it does maintenance. This activity is also compatible with other land use, such as farming for local markets and shellfish harvesting. The emerging concern for protecting the Cape's environment for the sake of its tourist industry, as reflected in the work of the Cape Cod Commission and the Association for the Protection of Cape Cod, indicates a growing awareness that the Cape needs to protect some of its historic and environmental past. Doing so may require Cape Codders to accept a slower trajectory into a more contained future.

The problems facing the Cape today are shared by peoples across New England and the globe. The Cape is a particularly fragile ecosystem. It is basically a sand spit sticking out into an increasingly violent Atlantic Ocean. Its eastern beaches are eroding at an alarming rate. Pollution dumped upon its surfaces quickly moves into its unified water system. But the Cape is not unique as a fragile ecosystem; it shares a similar condition with barrier islands across the eastern United States and the Gulf of Mexico. Building in areas vulnerable to ocean storms has been an American habit for the last hundred years. Overtaxed water systems and excessive nitrate loads have likewise plagued these coasts for over four generations. Increasing land prices along the Gulf and East coasts have changed the social and demographic mix, and the availability of the coast for recreation for those with limited resources is not just a Cape Cod problem.[42]

Closer to the Cape, tourism and vacation homes have become central parts of the economy for most of the New England coast extending from Maine, where homeowners are trying to stop locals from walking on paths along the coast that have been walked for generations, south to the shores of Long Island. Other coastal areas likewise are struggling with increased pollution loads, eroding beaches, rising prices, and tensions over jobs and the environment.[43] Nor are these only American problems as the people of coastal Spain can attest, where wealthy, mostly English, settlers have built extensively along a once open coast.[44]

Many of the problems facing the Cape in the regime of tourism are shared by communities not only across the New England coastline but also in its interior. The price of Vermont farmland is being pushed higher and higher by the state's popularity as a vacationland. As land values rise, the ability of Vermont farmers to continue to pursue their traditional activity is compromised, yet those very quaint Vermont farms are part of the state's appeal.

Communities across the nation have had to deal with similar intertwining and conflicted relationships—between local and larger (state and national) governing structures—that increasingly frustrate or help Cape Codders. The Cape's ambivalent relationship with the Cape Cod National Seashore, its dependence on state and federal funds and programs such as disaster relief or state subsidized insurance, and its integration into state and national environmental regulations all have parallels in the experiences of Americans from the California Redwoods to the Gulf of Mexico. It is not just external governing bodies that have come to play an ever-greater role in the lives of Cape Codders. The history of the Cape over its last 150 years has been a history of increasing integration into, and expanding influence of, larger regional and national economies and institutions. This also has been a reality for many American communities. The loss of social diversity in areas of natural beauty, as the wealthy have come to take up more and more of the land of those areas, is also not unique to the Cape. As the middle class of Cape Codders are finding it ever more difficult to maintain a place in the community, poor black and white farmers and fishermen across coastal South Carolina, Georgia, Florida, and the Gulf are losing a similar battle to hold their place in the new economy.

Although the problems of the Cape are not unique to the Cape but shared in different forms and shapes by communities and governing entities around the globe, so far few people have managed to find sustainable solutions. If Cape Cod can find a means of protecting access while maintaining the environment and the livelihoods of the local communities, perhaps it will have something very important to offer the rest of the world.

Henry David Thoreau may have been wrong when he predicted that for the "fashionable world" the Cape would "never be agreeable," but he was dead-on when he told us, "At every step we made an impression on the Cape, though we were not aware of it." In becoming more aware and in working to protect Cape Cod in its diversity, we can honor Thoreau and future generations of Cape Codders and those who will come to love it.

FURTHER READING

Because Cape Cod holds an iconic place in our national imagination, it has been the setting for significant works of both fiction and nonfiction. A complete bibliography of all the works pertinent to a study of the Cape would be a book in itself. What follows is a short annotated bibliography for the general reader interested in reading deeper into the Cape's past.

The best and most readable history is Josef Berger's WPA project, *Cape Cod Pilot* (1937). The other major standby is Henry C. Kittredge's *Cape Cod: Its People and Their History* (1930). Katharine Dos Passos and Edith Foley Shay's *Down Cape Cod* (1947), Joseph C. Lincoln's *Cape Cod Yesterdays* (1935) and *All Along Shore* (1931), Mary Rogers Bangs's *Old Cape Cod: The Land, the Men, the Sea* (1920), and Scott Corbett's *Cape Cod's Way: An Informal History of Cape Cod* (1955), all provide readable histories. More recently James C. O'Connell's *Becoming Cape Cod* (2003) is an excellent short history with photographs of the process by which the Cape sold itself as a resort area. Those looking for old photographs of the Cape should look to the Arcadia Press, for their Images of America series. Most Cape towns have their own book that Arcadia Press series. These books have wonderful pictures that capture the feel of the Cape in the nineteenth and early twentieth centuries. Among other things, one gets a dramatic visual image of the Cape without trees. Besides the Arcadia Press series, most Cape towns have histories written by local citizens. Many of these, such as Durant Echeverria's *A History of Billingsgate* (1991), are well researched and fascinating. An inquiry at the town library can easily bring forth these works, some of which are real joys to read.

Within a couple of generations after the initial white settlement, Cape Codders turned to the sea and fished. The history of the Cape is intimately wrapped up with the history of its fishing. Matthew McKenzie's *Clearing the Coastline: The Nineteenth-Century Ecological and Cultural Transformation of Cape Cod* (2010) is an excellent place to begin exploring Cape Codders' link to fishing. See also Daniel Pauly and Jay Maclean's, *In a Perfect Ocean: the State of Fisheries and Ecosystems in the North Atlantic Ocean* (2003). W. Jeffery Bolster's *The Mortal Sea: Fishing the Atlantic in the Age of Sail* (2012) is an excellent exploration of the role of marine resources in human history. For anyone interested in overfishing the North Atlantic, Cape Cod fishing communities, and the relationship between fisheries and fishing peoples and their knowledge of those fisheries, Bolster's book is a vital resource.

Interest in the Cape also produced memoirs and personal histories. The following books give readers a strong sense of Cape Cod in the nineteenth and early twentieth centuries from a variety of perspectives: Florence Baker, *Yesterdays Tide* (1876); Charles N. Barnard, *The Winter People: A Return to Cape Cod* (1973); Dennis and Marion Chatham, *Cape Coddities* (1920); George Howe Colt, *The Big House: A Century in the Life of an American Summer Home* (2003); Scott Corbett, *We Chose Cape Cod* (1953); Samuel Chamberlain, *Cape Cod in the Sun* (1937); Katherine Crosby, *Blue-Water Men and Other Cape Codders* (1946); Agnes Edwards, *Cape Cod New and Old* (1918); E. C. Janes, *I Remember Cape Cod* (1974); Anthony Marshall, *Truro, Cape Cod, As I Knew It* (1974); Joshua Atkins Nickerson, *Days to Remember* (1988); Edith Shay and Frank Shay, *Sand In Their Shoes: A Cape Cod Reader* (1951); Charles W. Swift, *The Right Arm of Massachusetts: An Historical Narrative* (1897); Gladys Taber, *My Own Cape Cod* (1971); and Arthur Wilson Tarbell, *Retire to Cape Cod* (1944).

Those interested in the Cape's history should also look at the three encyclopedic histories done of Barnstable County in the nineteenth and early twentieth century: Frederick Freeman's two-volume *The History of Cape Cod: The Annals of Barnstable County and Its Several Towns Including the District of Mashpee* (1862) has extensive documents concerning the early settlement of Cape Cod, including town minutes and colonial legislation and law. Simeon L. Deyo's *History of Barnstable County, Massachusetts* (1890) has not only rich detailed history of the Cape and its towns but also includes documents. Elroy Thompson's *History of Plymouth, Norfolk and Barnstable Counties, Massachusetts* (1928) brings the story up to the early twentieth century. Yale University's president Timothy Dwight (served 1795–1817), spent his summers touring for recreation and health. Those travels were published posthumously as *Travels in New England and New York* (1969). Volume III of Dwight's *Travels* covers Cape Cod. Edward Augustus

Kendall's *Travels through the Northern Parts of the United States in the Years 1807 and 1808* (1809) also has information on Cape Cod, although Kendall is not as reliable an informant as Dwight. He looked at the Cape through the narrow vision of an Englishmen who took the English countryside as his measure of what should be and did not appreciate Cape Codders' adaption to the local ecology. Those interested in how early Cape settlers dealt with the environment should look at Ruth Lynn Friedman's "Governing the Land: An Environmental history of Cape Cod, Massachusetts, 1600–1861" (PhD dissertation, Brandies University, 1992).

The Cape's natural history has drawn visitors since the early nineteenth century whose works have become classics in American naturalist writings. The most famous of these writers was Henry David Thoreau. His *Cape Cod* remains one of the most enjoyable and insightful natural histories of Cape Cod, although one should be careful to remember Thoreau had a tendency to say things tongue in cheek. Other books by Cape writers that should be on everyone's shelf include: Henry Beston's classic *The Outermost House* (1928); John Hay's *The Great Beach* (1963), *The Undiscovered Country* (1981), *Nature's Year: The Seasons of Cape Cod* (1961), and *The Run* (1959); Clare Leighton's, *Where the Land and Sea Meets the Sea: The Tide Line of Cape Cod* (1954); Walter Teller's *Cape Cod and the Islands* (1970); and more recently Robert Finch's *Special Places on Cape Cod and the Islands* (2003), *The Primal Place* (1983), and *Common Ground: A Naturalist's Guild to Cape Cod* (1981). Cynthia Huntington's, *The Salt House* (1999) is a nice description of life among the dunes of the Outer Cape.

For a more technical description of Cape Cod the following books are readable and reliable: *A Sierra Club Naturalist's Guide to the North Atlantic Coast, Cape Cod to Newfoundland* (1981), the Audubon Society's *Atlantic and Gulf Coasts: A Comprehensive Field Guild* (1985), and Stanley Schuler's *A Geologist's View of Cape Cod* (1982), Philip Kopper's *The Wild Edge* (1979), and Dorothy Sterling's *The Outer Lands* (1978). Also helpful is Kenneth L. Gosner's *A Field Guide to the Atlantic Seashore* (1978). On more specialized topics are Norman P. Hill's *The Birds of Cape Cod, Massachusetts* (1965) and Harold R. Hinds and Wilfred A. Hathaway's, *Wildflowers of Cape Cod* (1968).

The Nickerson Room (at Cape Cod Community College) and the Cape Cod National Seashore Visitor's Center in Eastham have extensive collections of Cape-related documents. A website about Cape Cod history and genealogy established in 2003 (capecodhistory.us) been updated with related documents, at the time of this writing, through March 2013. Anyone interested in the Cape's early history should check this site.

NOTES

INTRODUCTION

1. The Upper Cape refers to the towns clustered around its western end: Sandwich, Bourne, Falmouth, and Mashpee. It was called the Upper Cape because of the scales on maps and charts. To travel east was to go down the longitudinal scale toward England. To travel west was to go up the scale. Also, since prevailing winds were from the southwest, one traveled downwind to go east. The towns of Barnstable, Yarmouth, and Dennis were considered Mid-Cape by the end of the nineteenth century whereas the north–south section or Outer Cape, along with the close-by towns of Brewster and Harwich, came to be called the Lower Cape.
2. Charles Morrow Wilson, *Dow Baker and the Great Banana Fleet* (Harrisburg, PA: Stackpole Books, 1972), 111.
3. Simeon L. Deyo, *History of Barnstable County, Massachusetts* (New York: H. W. Blake, 1890), 791.
4. Wilson, *Dow Baker*, 60.
5. Cape Cod National Seashore, "Report on the Atwood Higgins House," 2009, Wellfleet Public Library.
6. Henry David Thoreau, *Cape Cod* (New York: Penguin Press, 1987; orig. 1865), 236.
7. Whitman and Howard, *What the Herring River Dike Project Would Mean to the Town of Wellfleet: Report of Messrs. Whitman and Howard on Proposed Dike at Herring River* (Wellfleet, MA: 1906), 12. The mosquito population ebbs and flows with spring rains. Heavy spring rains bring exceptionally large numbers of mosquitoes. Soon after Baker opened Chequessett Inn the Cape experienced heavy spring rains and a mosquito bloom. In Baker's mind this had to be controlled.
8. Baker hoped to get the state to pay half the cost of the dike, but that would still have forced the town to come up with $10,000.
9. Whitman and Howard, *What the Herring River Dike Project Would Mean*, 13.

10. An 1890 history of the Cape noted that the Herring River "has for many years been a source of profit. The annual run of herring to the ponds above to spawn yielded unfailing revenue." Deyo, *History of Barnstable County*, 630.
11. Dorothy Sterling, "Our Cape Cod Salt Marshes," Information Bulletin no. 6 (Orleans, MA: APCC, 1976), 12.
12. Thoreau, *Cape Cod*, 22. Thoreau's view of Cape Cod as a "weather-beaten garment" was typical of his larger vision of the world. Thoreau presented Concord in both his journals and his now-famous *Walden* as fallen from a previously purer state. Thoreau's four trips to Cape Cod were made in October 1849, June 1850, July 1855, and June 1857. He walked from Eastham to Provincetown three times, twice on the Atlantic side and once on the bay side, and he crossed the arm of the Cape some six times. Twice he made these walks with his friend William Ellery Channing and twice alone.
13. Cape Cod National Seashore Report, "Atwood Higgins House." Soil scientists estimated that 40 percent of Baker's Wellfleet had significant amounts of fertile soil in the seventeenth century, but by the beginning of the nineteenth century it was largely coarse culver sand and muck.
14. Pitch pine dominated the Cape forest 350 years ago as it does today. The periodic forest fires—both those set by Native Americans and those caused by lightning—advantaged pitch pine, whose thick bark is relatively fire resistant. The tree also has a dormant bud at its base that sprouts a new growth even if the trunk is burned. Although the Cape had significant forests of oak, beech, black gum, and hickory (as well as red maple and white cedar in the wetlands) before white settlement, the forest of today has more oak than that of 350 years ago. The pitch pine is not as shade-tolerant as oaks or even white pine. As oaks grow up and shade the pitch pine, these pines will give way to the oaks. Robert W. Eberhardt et al., "Conservation of Changing Landscapes: Vegetation and Land-Use History of Cape Cod National Seashore," *Ecological Applications* 13 (2003): 68–84.
15. The concept of regimes of resource utilization is similar but broader than Gadgil and Guha's concept of modes of resource utilization. See Madhav Gadgil and Ramachandra Guha, *This Fissured Land: An Ecological History of India* (Oxford: Oxford University Press, 1992). For an example of a work that uses a similar frame of reference see Mart A. Stewart, *"What Nature Suffers to Groe": Life, Labor, and Landscape on the Georgia Coast, 1680–1920* (Athens: University of Georgia Press, 1996).
16. William Cronon was one of the first historians to investigate the environmental history of New England. William Cronon, *Changes in the Land: Indians, Colonists and the Ecology of New England* (New York: Hill and Wang, 1983). Despite his nuanced look at the dynamic relationship between people and their environment, this early work tends toward a static view of native people's use of the region's resources before European contact. There is significant archeological evidence to indicate that sometime between 100 BCE and 900 CE the peoples on the Cape developed a resource use similar to the one Cronon depicts in his work of seasonal migrations, mobile agriculture, and seasons of hunger and of fullness. But by the time of European contact there were Cape communities, the culture of resource use allowed for year-round and more permanent settlements. For another early environmental history of New England, looking at dif-

NOTES TO PAGES 6–7   217

fering modes of production between Native Americans, colonial Americans, and, later, capitalist nineteenth-century Americans and how they impacted the environment, see Carolyn Merchant, *Ecological Revolutions: Nature, Gender, and Science in New England* (Chapel Hill: University of North Carolina Press, 1989).

17. The very land Henry David Thoreau traversed on some of his hikes across the Outer Cape is now over a hundred feet out to sea due to the action of storms, shifting sands, and rising sea level.

18. Cape Cod was not alone in losing population during the nineteenth century. Many rural communities saw their sons and daughters leave for greener fields to the west or better opportunities in the cities. While many communities lost populations the reasons for that loss varied tremendously across the Northeast. The hill towns of rural New England began losing people with the great cold years between 1814 and 1818. The completion of the Erie Canal hurt cereal producers across the region from the mid-1820s through the 1840s. Economic opportunity in the mills and shops in towns, cities, and ports pulled frustrated children of farmers away from much of rural New England. But each region and each farm had its own constellation of factors that accounted for its population loss. This book looks at the factors that played out across Cape Cod. Certainly the Cape shared some features with other areas that lost population, but it also had factors unique to itself, as reflected in the fact that over the first half of the nineteenth century, while much of rural New England was losing people, the Cape experienced strong population growth.

19. This point of change rather than decline is another point of distinction between this book and William Cronon's *Changes in the Land*. Cronon and Carolyn Merchant have done significant work in bringing the idea of changing modes of production into our understanding of environmental history. Their analysis of the importance of European concepts of property, the use of animals and European cropping, and commoditization of resources in changing the land has provided a framework all environmental historians of New England must address. Nonetheless *Changes in the Land* is not without problems. Although Cronon argues against the tendency to see the present as a fallen version of the past, he nonetheless posits the capitalist world of the late colonial period as an environmentally damaged place. This work sees more diversity and complexity in the changes that occurred between initial white contact and the present day. Whites brought concepts of commoditization and markets to the New World as well as disease, weapons, and animals. But the capitalism they brought was by no means fully hegemonic. Although markets, commoditization, and the search for profits pushed colonialists toward waste, it must also be remembered that these same colonists left behind an amazing history of resource husbandry. Capitalism has its destructive nature, but until the 1820s and 1830s, Cape Codders were somewhat successful in mitigating that destruction. And in the twentieth century we also see examples of both the destructive force of capitalism and the activity of citizens to mitigate that destruction. This mixed story is central to this book.

20. Mine is not the first environmental history set in New England to look at how different means of organizing an economy affect the environment in different ways. Besides

Cronon and Merchant, Diana Muir has written an interesting and very readable work on New England environmental history. Diana Muir, *Reflections in Bullough's Pond: Economy and Ecosystems in New England* (Hanover, NH: University Press of New England, 2000). See also Theodore Steinberg, *Nature Incorporated: Industrialization and the Waters of New England* (New York: Cambridge University Press, 1991); Brian Donahue, *The Great Marsh: Farmers and the Land in Colonial Concord* (New Haven: Yale University Press, 2007); and Richard Judd, *Common Lands and Common People: the Origins of Conservation in Northern New England* (Cambridge: Harvard University Press, 1997). This book adds to that literature by bringing the story up to the present. It also differs from much of recent work being done on New England environmental history, which has moved away from a materialist viewpoint toward a framework that emphasizes perception. This book embraces materialism and focuses directly on the interconnections between environmental history and economic history, while putting that within the larger framework of social history.

1. FROM CONTINENTAL DRIFT TO NOMADIC LAND USE

1. Scientists looking at the breakup of the large Arctic glaciers and those covering Greenland have argued that perhaps we are at the end of an epoch. Global climate change primarily initiated by human activity may well indicate the end of the Holocene epoch of relative climate stability. See Johan Rockstrom et al., "A Safe Operating Space for Humanity," *Nature* 461 (2009): 472–75.
2. Initially, as the glacier ice melted sea level rose dramatically, some 25 meters between 12,000 and 10,000 years ago. By 10,000 years ago most of the ice had melted back into the ocean basins and the sea level rise slowed to a rate of 6 meters every 1,000 years; 6,000 years ago the rise of sea levels slowed to 2 meters every 1,000 years. It reached its present level roughly 1,000 years ago. "Cape Cod Environmental Atlas," ed. Arthur B. Brownlow (Department of Geology, Boston University, 1979), paper located at the Cape Cod National Seashore, Wellfleet, MA. See also Robert N. Oldale, *Cape Cod and the Islands: The Geological Story* (Orleans, MA: Parnassus Imprints, 1992) and Oldale, "How Cape Cod and the Islands Were Formed," in *Guide to Nature on Cape Cod and the Islands,* ed. Greg O'Brien (Brattleboro, VT: Stephen Greene Press, 1990), 15–26.
3. See James W. Bradley, *Archeological Investigations at the Carns Site, Coast Guard Beach, Cape Cod National Seashore, Massachusetts,* Occasional Publications in Field Archeology, no. 3 (Northeast Region Archeology Program, National Parks Service, U.S. Department of the Interior, 2005) for a detailed analysis of the archeological evidence of Native American land use patterns on the Cape.
4. Kathleen Bragdon argues that with the introduction of maize and squash, shellfish became less important to the people of the Cape. Kathleen J. Bragdon, *Native Peoples of Southern New England, 1500–1650* (Norman: University of Oklahoma Press, 1996). Although corn, beans, and squash were indisputably important to the diet of the Cape peoples, shellfish remained central to their diet and provided the means by which they could establish permanent settlements.

5. Thomas Morton noted in 1637 that the Native Americans set fire to the woods regularly, often twice a year, to keep them open and clear. See L. Stanford Altpeter, "A History of the Forests of Cape Cod" (MS thesis, Harvard University, 1939), 9.
6. For a description of Native Americans before heavy white settlement see Daniel Gookin, *Historical Collections of the Indians of New England,* Massachusetts Historical Society Collections, vol. 1 (Boston, 1792).
7. The herring that ran up the Cape streams in late March or April were called alewives. They were a variety of the shad. They would stay in the ponds till summer and the young fry would go back to the sea in the fall to return three years later.
8. During colonial times settlers dug sluiceways between ponds to increase the spawning areas of these anadromous fish. Archeological evidence indicates that this practice stretched back well before white settlement. Native Americans were not only harvesting these fish, but also facilitating their abundance.
9. Henry David Thoreau, *Cape Cod* (New York: Penguin Books, 1987; orig. 1865), 138.
10. See Gookin, *Historical Collections of the Indians of New England,* 150.
11. David W. Black and Ruth Holmes Whitehead, "Prehistoric Shellfish Preservation and Storage on the Northeast Coast," *North American Archeologist* 9 (1988): 17–30.

2. FIRE, FISHING, AND FARMING OF NATIVE PEOPLES

1. Quoted in Florence Wing Baker, *Yesterday's Tide* (privately printed, 1941), 5.
2. Ibid., 6, 7.
3. For a description of the native community on the Outer Cape see Durant Echeverria, *A History of Billingsgate* (Wellfleet, MA: Wellfleet Historical Society, 1993), 12.
4. Originally the native tribes on the Cape (the Pamets of the Northern Cape, the Punonakanits of Wellfleet and Eastham, the Nausets of Orleans, the Monomoyicks of Chatham, the Mattakeese of Yarmouth, the Cummaquids of Hyannis, the Monomets of Sandwich, and the Sakonnets of the Falmouth area) were loosely allied with the Pokanokets, the tribe of the Pilgrims' friend Massasoit. The epidemic that swept through New England native communities in 1616–1619 hit the Cape native communities with less severity than the villages on the mainland, but following the attack of the Pilgrims against the natives south of Boston in 1623, the Cape Native Americans fled their villages and stocks of food in fear that they would be attacked. They took refuge in the swamps of the Cape, where hunger and disease hit them as the plague of 1616–1619 had not. Soon all the leading Cape sachems were dead. Following the collapse of Cape Indian leadership, Massasoit molded the groups into the Wampanoag tribe. Nathaniel Philbrick, *Mayflower: A Story of Courage, Community and War* (New York: Viking Press, 2006), 154–55. See also Bragdon, *Native Peoples.*
5. According to Gookin, the Native Americans boiled the maize with the beans and squash. They would add to this fish, clams, or any meat they had available. They would often thicken the mixture with groundnuts. Gookin, *Historical Collections of the Indians of New England,* 150. In 1991, the anthropologist Stephen Mrozowski excavated fish bones from corn hills of a Wampanoag cornfield, dating to before white settlement.

See H. Bruce Franklin, *The Most Important Fish in the Sea: Menhaden and America* (Washington, DC: Island Press, 2007), 14.
6. It was just such a cache of food that the Pilgrims pillaged when they first landed in Provincetown before they crossed over to Plymouth.
7. Altpeter, "A History of the Forests of Cape Cod," 12, 13.
8. Because of the importance of shellfish and coastal plants to the native diet, almost all the native settlements were along the coast while the interior of the Cape was mostly left for hunting or trails to other settlements. The exceptions to this were settlements along the larger ponds such as Mill Pond in Brewster, or along the tidal rivers such as the Bass River, which runs two-thirds the way up from Nantucket Sound into a series of large brackish ponds.
9. For a general description of the history of Native Americans in Massachusetts see Jack Tager, Richard W. Wilkie, Dena F. Dinauze, and Mitchell T. Mulholland, "Native Settlements and European Contacts," in *Historical Atlas of Massachusetts*, ed. Richard W. Wilkie and Jack Tager (Amherst: University of Massachusetts Press, 1991), 10–15. For an interesting and very readable description of native peoples of New England and their interactions with whites see Diana Muir, *Reflections in Bullough's Pond: Economy and Ecosystems in New England* (Hanover, NH: University Press of New England, 2000).
10. Thoreau, *Cape Cod*, 314.
11. Altpeter estimates that in 1620 only 1.5 percent of the Cape was open corn land, and another 2 percent may have been abandoned cornfields returning to pine forests. Forests affected by native burning may have been as great as 61 percent of the land, whereas 32 percent of the land would have been mesophytic (moderately moist). Altpeter, "A History of the Forests of Cape Cod," 12.
12. There is considerable speculation that the Vikings were the first Europeans to visit Cape Cod. Since the evidence of such a visit is slim and it makes no difference to this work, I will leave that question to others.
13. The epidemic that swept through the Cape communities probably struck in 1612–1613 and returned in 1616 and 1619. The white settlers who arrived a few years after the plague considered it an act of God. "Thereby divine providence made way for the quiet and peaceable settlement of the English." Gookin, *Historical Collections of the Indians of New England*, 148. For a reference to trichinosis as a possible source of the plague in this period, see "New Hypothesis for Cause of Epidemic Among Native Americans, New England, 1616–1619," *Emerging Infectious Diseases* 16 (February 2010): 281–86.
14. At Plymouth, where Champlain in 1605 had found several wigwams, fields of corn, beans, and squash, as well as canoes and open land, the Pilgrims in 1620 found the region full of human skulls and bones but empty of people.
15. The Pilgrims had English backers who expected a return on their investment. Moreover, their agricultural methods, unlike those of the Native Americans, required tools and animals that had to be purchased in England and brought over. This pushed the Pilgrim settlers into a desperate search for a commodity valued in England that would allow them to pay for their needed tools and animals as well as to help satisfy their

English creditors. Furs, particularly beaver furs, provided that commodity. To get furs the settlers traded goods to the Native Americans.
16. Henry C. Kittredge, *Cape Cod: Its People and Their History* (Boston: Houghton Mifflin, 1930), 28–49.

3. ON THE WAY TO AN AMPHIBIOUS SOCIETY

1. Quoted in L. Stanford Altpeter, "A History of the Forests of Cape Cod" (MS thesis, Harvard University, 1939), 6. It is unclear if these were true species of trees, since the species in England were not the same as those on this side of the Atlantic. Much of the underwood was burnt back by Native Americans as part of their land management practices. When whites settled the region they abandoned the practice of burning the understory. This resulted in a considerable buildup of brush, leading the town of Sandwich to set up a committee of 42 men in the mid-eighteenth century to "fire the woods before April 16th." Simeon L. Deyo, *History of Barnstable County, Massachusetts* (New York: H. W. Blake, 1890), 273.
2. See Alfred Crosby, *Ecological Imperialism: The Biological Expansion of Europe, 900–1900* (New York: Cambridge University Press, 1986) for a discussion of this process across the whole of the New World.
3. The hurricane of 1635 is considered by scientists to have produced the largest storm surge in the last four hundred years.
4. See Brian Donahue's excellent *The Great Meadow: Farmers and the Land in Colonial Concord* (New Haven: Yale University Press, 2004) for a discussion of the role of meadow grasses in sustaining winter animals in New England.
5. The original settlers of Plymouth had been awarded on average 20 acres per family.
6. Quoted in Josef Berger, *Cape Cod Pilot* (Cambridge: MIT Press, 1969), 161.
7. Altpeter, "A History of the Forests of Cape Cod," 6.
8. The Native American practice of setting fire to the woods—as well as the intensely hot summer sun just inland from the coast, especially on the Outer Cape, which dried out the land and made it more susceptible to lightening fires—encouraged the growth of pitch pine more so than deciduous trees. Indeed the high concentrations of Native Americans, combined with more flammable conditions, led to more frequent fires than in inland New England. W. A. Patterson III, E. K. Saunders, and L. J. Horton, "Fire Regimes of Cape Cod National Seashore," U.S. Department of Interior, National Park Service, Report OSS 83-1 (Boston, 1984); T. Parshall, D. R. Foster, E. Faison, D. MacDonald, B. C. S. Hansen, "Long-Term Vegetation and Fire Dynamics of Pitch Pine-Oak Forests on Cape Cod, Massachusetts," *Ecology* 84 (2002): 736–48. In areas where oaks can establish a foothold under the pines, oaks (as long as they were not killed by fire) will over-canopy the pines, which are not shade-tolerant, and replace them. Within the oak forests other deciduous trees such as beech and hickory grew as well. This helps account for the diversity of the Cape's forests. Where there was regular native firing of the forests, or where the ground cover dried out sufficiently to allow for lightning fires, pitch pine and then white pine prevailed. Where there was less firing

of the forest, and the groundcover more shaded and moist, deciduous trees established themselves and pushed out the conifers. In the wet marshy depressions, white cedar and red maple took hold. J. W. Blankinship, "The Plant Formations of Eastern Massachusetts," *Rhodora* 5 (1903): 124–37. See also S. Little, "Fire and Plant Succession in the New Jersey Pine Barrens," in R. T. T. Forman, ed., *Pine Barrens: Ecosystems and Landscape* (New York: Academic Press, 1979), 297–314.

9. Altpeter, "A History of the Forests of Cape Cod," 21.
10. Deyo, *History of Barnstable County,* 266.
11. Quoted ibid., 268.
12. Much work has been done on the role of the commons for early settlements. See, for example, Michael Rawson, *Eden on the Charles: The Making of Boston* (Cambridge: Harvard University Press, 2010).
13. Deyo, *History of Barnstable County,* 268.
14. Ibid., 269. Dexter did not abandon his hopes. A few years later, he built a mill and a dam (it remains standing to this day) under an agreement that the town would pay him five pints for each bushel ground. The town later objected to the high toll he charged and built a town-owned mill at Spring Hill in 1668. Facing the same problem but lacking streams near the original settlement, Barnstable in 1687 ordered that a windmill be built. The mill was built on Meeting House Hill and ground town grain for several years.
15. Quoted in Elroy Thompson, *History of Plymouth, Norfolk and Barnstable Counties, Massachusetts* (New York: Lewis Publishing Co., 1928), vol. 2, 730. At the time of colonial settlement moose, deer, wildcats, beaver, fox, skunks, muskrats, rabbits, woodchucks, wild turkeys, and heath hens were plentiful along with sea and shore birds.
16. The nineteenth-century historian John Gorham Palfrey, originally from Barnstable, noted that the colonists purchased the land from the Native Americans on the cheap. He attributed this to the Native Americans' "idle and roving habits." This may have eased the consciences of the whites, but it did damage to the historical reality of permanent native settlements and their agricultural activity. Palfrey, quoted in Frederick Freeman, *The History of Cape Cod: The Annals of Barnstable County and Its Several Towns Including the District of Mashpee* (Boston: George Rand and Avery, 1862), vol. 2, 258. See also Deyo, *History of Barnstable County,* 369.
17. Deyo, *History of Barnstable County,* 370.
18. Herman Melville, *Moby Dick* (New York: Alfred A. Knopf, 1991), 52.
19. Freeman, *The History of Cape Cod,* vol. 2, 358. The animals were earmarked with cuts to distinguish ownership. Thoreau noted that this practice continued into the nineteenth century with "many horses . . . still turned out to pasture all summer on the islands and beaches in Wellfleet, Eastham and Orleans as a kind of common." Thoreau, *Cape Cod* (New York: Penguin Press, 1987), 103.
20. This marsh hay continued to be important for Cape farmers into the late nineteenth century. In 1783 the minister of Chatham was allotted 18 cords of wood, but also 4 loads of salt hay for his livestock. In 1796, while his allotment of wood dropped to 16 cords, his salt hay allotment grew to 5 loads. Freeman, *The History of Cape Cod,* vol. 2, 606.

21. Thompson, *History of Plymouth, Norfolk and Barnstable Counties,* vol. 2, 730, 757; Freeman, *The History of Cape Cod,* vol. 2, 374. Birds were not the only indigenous fauna Cape Cod farmers attempted to eliminate. Wolves preyed upon sheep and young cattle that farmers let wander on the outer dune grasses. Town fathers paid a bounty for anyone killing a wolf. In 1655 bounty was paid for six slain wolves and bounties continued to be paid through the eighteenth century. In 1789 Barnstable offered $50 for a wolf killed in town and $25 for one killed outside of town. Freeman, *The History of Cape Cod,* vol. 2, 187.
22. Berger, *Cape Cod Pilot,* 179.
23. Altpeter estimated that a typical Cape home contained 15,000–20,000 board feet of wood (he may have been including the barn). The thousand or so farms that dotted the Cape's landscape in the late 1600s would have consumed over 30 million board feet of lumber. Altpeter, "A History of the Forests of Cape Cod," 24.
24. Katherine Crosby, *Blue-Water Men and Other Cape Codders* (New York: Macmillan, 1946), 63.
25. Timothy Dwight, *Travels in New England and New York,* vol. 3, ed. Barbara Miller Solomon and Patricia King (Cambridge: Harvard University Press, 1969), 51. These Cape cottages were initially shingled in pine shakes, but by the eighteenth century the good white pine used for shingles was so depleted that homebuilders shifted to cedar shakes or shingles. The cost of one of these homes in the eighteenth century, using local timber, could be about $200 to $300. Katharine Dos Passos and Edith Foley Shay, *Down Cape Cod* (Boston: Robert M. McBride, 1947), 41.
26. Altpeter, "A History of the Forests of Cape Cod," 58; Freeman, *The History of Cape Cod,* vol. 2, 600. By the mid-eighteenth century wood had become scarce and the allotments of wood dropped to 18 to 20 cords of wood. The Reverend Caleb Upham, when appointed to his post in Truro in 1755, was allotted 20 cords of wood delivered to his home for heat and cooking each year. Freeman, 558. The allotment in Chatham dropped from 30 cords to 18 by the middle of the eighteenth century. Freeman, 605.
27. Freeman, *The History of Cape Cod,* vol. 2, 606.
28. Ibid, 192. When land was divided among the original settlers, or when later land divisions were made, consideration was given to these needs. In the land distribution in Yarmouth in 1672 the allocations included "a parcel of land for a house-lot . . . and land on the east side of the river containing ten acres and a parcel of meadow land . . . containing 6 acres and another parcel . . . in the meadow at Canoe Point and another of planting land at Nobscussetts containing 8 acres."
29. The Cape's wheat crop was hit by blight in 1665, which caused farmers to give up sowing wheat for several years. But by the end of the century wheat was again a major crop.
30. By 1686 Barnstable farmer John Howland was producing enough surplus cider that the town allowed him to "retail his cider." Quoted in Freeman, *The History of Cape Cod,* vol. 2, 277.
31. Altpeter, "A History of the Forests of Cape Cod," 22, 23.
32. Freeman, *The History of Cape Cod,* vol. 2, 23, 30.

33. Ibid., 554. This issue continued to haunt town meetings on the Cape throughout the eighteenth century.
34. Ibid., 297.
35. Millers usually took about 5 percent of what was ground in payment for grinding.
36. To this day Wellfleet and Eastham, taken together, have the third-largest number of stranding of *Cetacea* in the world.
37. In 1702 the town of Sandwich gave its minister the right to any whales that drifted ashore as pay for his ministry. Eastham deemed all whales coming ashore to be the property of the town for the benefit of the minister. Thompson, *History of Plymouth, Norfolk and Barnstable Counties*, vol. 2, 730.
38. Freeman, *The History of Cape Cod*, vol. 2, 558.
39. Ibid.
40. Deyo, *History of Barnstable County*, 135.
41. Boat builders used local oak and pine to fashion their boats. By the end of the eighteenth century settlers were planting black locust, brought up from the southern colonies. Black locust is a hard, dense wood that is disease and pest resistant, and it grew easily and fast in Cape soil. Initially farmers were the ones interested in black locust for fence posts and foundation plates of their homes and barns, but soon boat builders were also utilizing the locust for frames for their boats.
42. Ship tonnage is not a weight measurement but a volume measurement. It indicates how many barrels a ship can carry.
43. Ship and tackle also needed turpentine and pine sap for caulking and preserving rope and wood. Turpentine and resin were distilled from pinesap, which came from tapping the trees, either by boxing or barking. Done carefully the trees could sustain this practice, but sustained harvesting was a slow process. Done hurried or carelessly it killed the tree.
44. Altpeter, "A History of the Forests of Cape Cod," 39.
45. Dos Passos and Shay, *Down Cape Cod*, 13.
46. A nineteenth-century quintal was a long hundredweight (cwt), or 112 pounds.
47. Freeman, *The History of Cape Cod*, vol. 1, 619.
48. Ibid.
49. Quoted in Altpeter, "A History of the Forests of Cape Cod," 33.
50. Quoted in Freeman, *The History of Cape Cod*, vol. 1, 349. In 1729 Provincetown was incorporated and the town's people were allowed to cut and take any trees necessary for building wharves, stages, homes, and flakes, but they could not "spoil, waste, bark, or box any standing pine. Ibid., vol. 1, 363.
51. Quoted ibid., vol. 1, 369–70. Throughout the eighteenth century the General Court kept passing legislation prohibiting or limiting the grazing of cattle and the cutting of trees, except for brush "necessary for flakes in the curing of fish" or, as in 1786, the 20 cords for the minister. Quoted ibid., vol. 1, 552. See also 375 and 382.
52. Ibid., vol. 2, 510.
53. Ibid., vol. 1, 323.
54. In 1714, for example, the Massachusetts General Court passed an act "regulating the inhabitants and sojourners [of Provincetown]. "The act noted that the harbor was being

"made wholly unserviceable by destroying the trees on the said cape," and among other things prohibited any act to "bark or box any pine trees . . . on the said cape for the drawing of turpentine." Freeman, *The History of Cape Cod*, vol. 1, 349.
55. Quoted in Durant Echeverria, *A History of Billingsgate* (Wellfleet, MA: Wellfleet Historical Society, 1993), 19.
56. Ruth Lynn Friedman, "Governing the Land: An Environmental History of Cape Cod, Massachusetts, 1600–1861" (PhD dissertation, Brandeis University, 1992), 107, 119.
57. Ibid., 162. Yarmouth was not the only town asking for help in protecting its fragile lands. Provincetown, Truro, Eastham, Barnstable, and Sandwich also attempted to pass ordinances restricting land use throughout the seventeenth and early eighteenth century as well as turning to the General Court (first in Plymouth, then in Boston) for colonial legislation to protect the land and sea resources. In 1767, Wellfleet petitioned the General Court for protection of land from overgrazing by "cattle, horsekind and sheep." Quoted in Freeman, *The History of Cape Cod*, vol. 1, 172. For other towns attempt to restrict land use and their petitions to the general court for aid in restricting land use see Freeman, *The History of Cape Cod*, vol. 1, 119–80.
58. Even amid the struggle for independence the 1770s saw attempts to restrict land use on the Cape. In 1774 the General Court restricted the number of livestock that could be grazed on the lands around Provincetown to three yoke oxen for the whole town and one horse and one cow per family. It also prohibited the cutting down of trees growing within "160 poles from the high-water mark." Quoted in Freeman, *The History of Cape Cod*, vol. 1, 382. In 1786 the cutting of brush and trees was prohibited except for the 20 cords for the minister and as much brush as necessary for flakes. Ibid., 552. See also Echeverria, *A History of Billingsgate*, 109.
59. Massachusetts Historical Society, *Collections of the Massachusetts Historical Society, 1794*, vol. 3 (Boston: 1794), 14.
60. Timothy Dwight, *Travels in New England and New York*, vol. 8, 57.
61. James Freeman, *A Description of the Eastern Coast of the County of Barnstable from Cape Cod Harbor or Race Point to Sandy Point of Chatham* (Boston: H. Sprague, 1802), 5, 9. This manuscript is located in the Nickerson Room, Cape Cod Community College Library.
62. Freeman, *The History of Cape Cod*, vol. 1, 627. Not all Cape Codders saw the future and human relationship to the land the same way. For Native Americans it became increasingly clear their ancestors who had greeted the Pilgrims with arrows had the correct approach. Throughout the eighteenth century, Cape Indians petitioned the General Court protesting English taking of their land and building houses on it, grazing their animals on Indian lands, and cutting Indian wood and marsh land without compensation: "Englishmen use our lands . . . they cut our woods." Throughout most of the century the General Court ignored these petitions, but finally in 1763 it incorporated Mashpee as a special town for the Native Americans. Mashpee was a mixed blessing for the Native Americans, it gave them land that was more protected from white use but the General Court put far more restrictions on Mashpee than on any of the other towns. See Friedman, "Governing the Land," 151–55.

63. Wellfleet limited the time when oysters could be harvested and finally in the 1770s restricted oyster harvesting to local families—not for export—and only allowed their harvest by hand, not by rake or other instrument. Friedman, "Governing the Land," 174, 175.
64. In 1671 Plymouth passed regulations prohibiting the taking of fish except at specific times. Again in 1678 the colony restricted fishing in colonial waters to only its and Boston's fishermen. In 1684 the court prohibited the catching of mackerel with nets or seines. In 1691 the colony put the regulation of fishing in the hands of Barnstable County, but in 1694 the new Massachusetts government again began regulating fishing and restricting how and when fish could be caught. Freeman, *The History of Cape Cod,* vol. 1, 266, 268, 299, 318. In the mid-eighteenth century the residents of Wellfleet petitioned the General Court, which had already outlawed grazing of fragile beach meadows, for additional help because of a decline of near-shore "whale fishing" and because fishing "has failed of late." Quoted in Friedman, "Governing the Land," 139.
65. Friedman, "Governing the Land," 172.
66. Ibid., 156, 157, 158.
67. In frustration with its inability to control overcutting of wood on its commons the town of Eastham decided in 1711 to sell off its common land in hopes private owners would be better able to protect the woodlands. Unfortunately, the new owners bought the land for the lumber and quickly the land was completely lumbered. Echeverria, *A History of Billingsgate,* 102. See Friedman's excellent dissertation (particularly pages 119–80) for a detailed description of the attempts of these early towns and the colonial government to address the issue of environmental degradation.
68. Deyo, *History of Barnstable County,* 132.
69. During the Revolution a bounty of 3 shillings a bushel for salt manufacture was established, which encouraged the improvement and expansion of solar evaporation for salt production. The new solar-evaporation method, first developed in Dennis in 1776 by John Sears, involved windmills pumping salt water into wooden vats that lined the shore. These vats had retractable covers that would be removed from the vats in sunny weather but would cover them during rain. The evermore salt-concentrated water would be pumped into new vats as it became heavy with salts, until it would crystallize in the final vats. Solar evaporation reduced the need for wood fuel, but the wooden vats required a massive amount of lumber.
70. Initially travel on and off the Cape involved linking up with whatever captain was carrying goods (usually wood, fish, or grain) to the desired destination. Sandwich established a regular packet service to Boston in 1717. By the beginning of the nineteenth century packet services were common. Until the late 1840s these services used sailing schooners to carry goods and passengers to Plymouth, Boston, and points north or New Bedford, Fall River, Providence, or New York. These packet boats carried between 25 and 50 people crowded into cabins and on deck. Until the railroad these were the main means of moving people and goods on- and off-Cape.
71. Massachusetts Historical Society, *Collections, 1794,* vol. 3, 14.
72. Ibid.

73. Ibid.
74. Ibid., 199.
75. Ibid.

4. MINING THE BOUNTY OF NATURE

1. Timothy Dwight, *Travels in New England and New York,* vol. 3, 48. Barbara Miller Solomon and Patricia King ed. (Cambridge: Harvard University Press, 1969), 48. Wendell Davis described in 1802 how the meadows and marshes of the Cape were "a great source of wealth and improving husbandry. By means of them they are enabled to keep large stocks of cattle in the winter and food for their subsistence through the remainder of the year, if necessary. It is computed that about 100 loads of salt hay are annually sold [for export]." Quoted in Massachusetts Historical Society, *Collections of the Massachusetts Historical Society* (Boston: 1802).
2. Dwight, *Travels,* 51. Dwight noted that the soil between Ware and Sandwich was poor and hardly fit for crops.
3. Ibid., 49.
4. Ibid., 58.
5. Ibid., 57. According to Josef Berger, by feeding their cows on salt hay, Eastham farmers were sending milk to the mainland well into the middle of the nineteenth century. Josef Berger (Jeremiah Digges), *Cape Cod Pilot* (Cambridge, MA: MIT Press, 1969; orig. 1937), 89. 158. Dwight (*Travels,* 57) noted that Eastham was sending over 1,000 bushels of corn annually to external markets.
6. Dwight, *Travels,* 51.
7. Quoted in Frederick Freeman, *The History of Cape Cod: The Annals of Barnstable County and of Its Several Towns Including the District of Mashpee* (Boston: George Rand and Avery, 1862), vol. 2, 716.
8. Quoted ibid., vol. 2, 718.
9. Dwight, *Travels,* 58.
10. Ibid., 63. Dwight found Provincetown fishers were catching over 37,000 quintals of cod, and 5,000 barrels of herring annually with a quintal of cod selling for $3.30 and a barrel of herring bringing in $4.00. Cod was measured in quintals with each nineteenth-nineteenth-century quintal equal to roughly a long hundredweight (cwt), which translates today into 112 pounds of fish. Mackerel and herring were measured in barrels.
11. Massachusetts Historical Society, *Collections, 1802,* vol. 8 (Boston, 1802), 196.
12. This represented some $25,137 in value of cod and $25,762 of mackerel. The Dennis fishing community invested $29,682 in its fisheries and employed some 274 men. Freeman, *The History of Cape Cod,* vol. 2, 712.
13. Ibid., vol. 2, 178.
14. Dwight, *Travels,* 71.
15. Freeman, *The History of Cape Cod,* vol. 2, 701.
16. Quoted ibid., vol. 2, 334.

17. Ibid., vol. 2, 538.
18. Durant Echeverria, *A History of Billingsgate* (Wellfleet, MA: Wellfleet Historical Society, 1993), 95.
19. Massachusetts Historical Society, *Collections, 1802*, 17.
20. Simeon Deyo, *History of Barnstable County, Massachusetts* (New York: H. W. Blake, 1890), 838.
21. Massachusetts Historical Society, *Collections, 1802*, 196.
22. Ibid., 196. In the 1970s a Basque whaling boat, the *San Juan*, was excavated off the coast of Labrador. The vessel had set sail from Basque Country and eventually sank in Red Bay in 1565. The Basques set up some 20 shore-based tryworks that have been excavated so far, around the shore of Red Bay. In the sixteenth century before the Pilgrims arrived in Plymouth it is estimated that the Basques had annual fleets of up to 30 ships employing over 2,000 men producing whale oil for European markets.
23. Agnes Edwards, *Cape Cod New and Old* (Boston: Houghton Mifflin, 1918), 128.
24. Freeman noted in 1862 that although whales were once in abundance, by the mid-nineteenth century they were scarce in the surrounding waters. Freeman, *The History of Cape Cod*, vol. 2, 623. By 1800 the five whaling boats out of Wellfleet that looked for whales off Newfoundland or the Strait of Belle Isle all carried salt and lines for catching cod. Freeman, 678.
25. The best description of nineteenth-century whaling still has to be Herman Melville's *Moby Dick*. Melville provides the reader with a detailed description not only of the dangers of whaling, but the costs as well.
26. Katherine Crosby, *Blue-Water Men and Other Cape Codders* (New York: Macmillan, 1946), 250, 251. Wellfleet's whaling boats were built in "her own yards of her own timber." Edwards, *Cape Cod New and Old*, 128.
27. Grand Banks was but one of a series of banks stretching north along the east coast of Nova Scotia and off southern Newfoundland. Besides the Grand Bank, Cape fishermen fished Browns Bank, Sable Island Bank, Baquereau Bank, and St. Pierre Bank.
28. Freeman, *The History of Cape Cod*, vol. 2, 573.
29. See Mark Kurlansky, *Cod: A Biography of the Fish that Changed the World* (New York: Walker & Company, 1997) for a description of the importance of cod in world history.
30. See Matthew McKenzie, *Clearing the Coastline: Nineteenth-Century Ecological and Cultural Transformation of Cape Cod* (Hanover, NH: University Press of New England, 2010), 61–63, for an excellent description of the changing technology of cod fishing in the nineteenth century.
31. The importance of herring for bait is reflected in the fact that when Wellfleet restricted the catching of river herring to three days a week in 1773 they accepted the catching of the fish for codfish bait.
32. Colonial and early-nineteenth-century New Englanders did not recognize menhaden, river herring, and alewives as separate species of fish from herring; they are all in the herring family. Unlike true herring, shad, alewives, and river herring (all of which consume zooplankton), menhaden (*Brevoortia tyrannus*, being the best type of bait fish), consume phytoplankton (mostly algae). Both groups of fish are planktivores that

take in food by filtering out the plankton, and neither can be caught with a baited hook. Because menhaden have no competition for the phytoplankton they eat, and because phytoplankton is abundant in the ocean, menhaden are the most prolific of our ocean fish. They travel in huge schools and are easily trapped. Alewives, shad, and river herring are anadromous and spawn in fresh water. Each spring thousands of these fish migrated up Cape streams and rivers to spawning grounds.

33. Henry David Thoreau, *Cape Cod* (New York: Penguin Press, 1987), 138. Although Thoreau wrote of the sea's "inexhaustible fertility" in 1849, the supplies of inshore fish were in fact declining. See Mathew McKenzie's thorough study of the Cape's inshore fishing industry for a discussion of the changing nature of inshore fish stock in response to natural changes in the coastal ecology and human extraction. McKenzie, *Clearing the Coastline*.

34. Freeman, *The History of Cape Cod*, vol. 2, 402.

35. Thoreau, *Cape Cod*, 40. Thoreau reported that locals believed that because clams were so plentiful they needed to be harvested out every two years to make room for more.

36. See John T. Cumbler, *Reasonable Use: The People, the Environment and the State* (Oxford: Oxford University Press, 2001) for a discussion of the importance of anadromous fish.

37. Freeman, *The History of Cape Cod*, vol. 2, 663.

38. Ruth Lynn Friedman, "Governing the Land: An Environmental History of Cape Cod, Massachusetts, 1600–1861" (PhD thesis, Brandeis University, 1992), 157. Deyo, *History of Barnstable County*, 273. Protection of spring fish runs continued through the early national period across the Cape. The General Court stepped in to protect alewives in Falmouth in 1798 and for Chatham, Harwich, and Yarmouth in 1813, 1814, and 1815, respectively. Freeman, *The History of Cape Cod*, vol. 1, 577, 605.

39. Frederick Freedman quotes state legislation as well as petitions from towns for restrictions on catching migratory fish and for laws mandating the opening of sluiceways so that the fish could get beyond dam restrictions. See Freeman, *The History of Cape Cod*, vol. 1, 552, 577, 612.

40. The sluice way between Gull Pond and Higgins that is so popular among summer canoers and kayakers was originally dug out to provide access to Gull Pond for the river herring in hopes of increasing spawning grounds and encouraging the migration of even more fish.

41. Joseph Crosby Lincoln, *Cape Cod Yesterdays* (Boston: Little Brown, 1935), 83.

42. Freedman believed that the oysters were "nearly exterminated" along the Cape shores by overharvesting. Freeman, *The History of Cape Cod*, vol. 2, 32.

43. Formal farming of oysters did not begin in Wellfleet until the beginning of the twentieth century when the state allowed the town to lease offshore water to shellfish harvesters in 1911.

44. Berger, *Cape Cod Pilot*, 89.

45. The dramatic increase in economic activity around the harbors in the early nineteenth century is reflected in the increased property damage done by storms. In 1723 a horrific Northeaster tore through the Cape with waters rising three to four feet above the previous high-water mark. Yet little damage was done by this storm since most of

the Cape's population and structures were in from the coast. Even the storm of 1770 had done significantly less damage than the hurricane of 1815, although it was just as fierce. By the time of the 1815 hurricane substantial development had accrued along the coast, wharves, warehouses, shops, saltworks, and boat works. This time the storm did massive amounts of damage to homes, wharves, saltworks, and boats. Freeman, *The History of Cape Cod*, vol. 2, 607.

46. Freeman, *The History of Cape Cod*, vol. 2, 692.
47. The central wharf was built to accommodate the increased activity at Duck Creek because the older Wellfleet harbors at Bound Brook and Duck Harbor were silting in and had to be abandoned. The silting of the these harbors also led to the gradual abandonment of homes on Bound Brook and Griffen Island, some of which were moved into the new town center. In 1870 Wellfleet built its last big wharf, the Mercantile Wharf, which included not only the wharf but a store as well, owned by the same company that built the wharf. Unfortunately for these investors, although they optimistically enlarged their wharf in 1883, the combination of the coming of the railroad and the decline in the fishing trade meant that their visions for profits did not materialize. Deyo, *History of Barnstable County*, 805.
48. John Braginton Smith and Duncan Oliver, "Port on the Bay, Yarmouth's Maritime History on the North Sea, 1638–the present" (Yarmouth, Historical Society of Old Yarmouth), Cape Cod Community College Library, Nickerson Rm. 14.
49. Massachusetts Historical Society, *Collections, 1794*, 16.
50. Deyo, *History of Barnstable County*, 530.
51. Ibid., 111.
52. Ibid., 120.
53. Marion Crowell Ryder, *Cape Cod Remembrances* (Taunton, MA: Dennis Historical Society, 1972), 51.
54. Deyo, *History of Barnstable County*, 528.
55. Squid and herring as well as mackerel were also caught in weirs that were set up offshore. These weirs acted as traps drawing the fish into circular pens. The fishermen would then sail out to their weirs and scoop up the herring or squid. Until the 1850s mackerel were mostly taken by hook and line with fishermen getting paid a share of the catch. When purse seine netting was developed for mackerel fishing in 1853 it quickly replaced line fishing. It also involved a heavier capital investment. Purse seine fishermen were paid in wages, not shares. Many complained that they realized far less take home pay under the wage system than the older share system. Deyo, *History of Barnstable County*, 135, 136.
56. Freeman, *The History of Cape Cod*, vol. 2, 692.
57. Ibid., 692.
58. Deyo, *History of Barnstable County* 737.
59. Today Bound Brook is landlocked. The inlet at Duck Harbor sanded over completely in the late nineteenth century through the natural process of migrating sand moving down from Provincetown.
60. Smith and Oliver, "Port on the Bay.

61. Deyo, *History of Barnstable County*, 142.
62. The large clipper ships tended to be built on the Cape, but rigged in Boston.
63. The process of taking plans and converting them into useable patterns either for boats or sails was called lofting, mostly likely because it was done in the loft where there was plenty of open floor space.
64. Dennis, for example, sent 18 vessels with a total tonnage of 1,037 out for cod and mackerel in 1837. These fisheries used over 16,691 bushels of salt. Freeman, *The History of Cape Cod*, vol. 2, 712.
65. Dwight, *Travels*, 51.
66. Massachusetts Historical Society, *Collections, 1802*, 124. By 1810 Massachusetts was producing 150,000 bushels of salt, the overwhelming majority of which was produced on the Cape.
67. Sea salt, once the water has been removed, is basic sodium chloride (NaCl), formed by a chemical reaction between an acid and a base. Glauber's and Epsom salts are hydrates, which include water in their solid crystal form.
68. Frederick Freeman in 1862 noted that although there were few saltworks left in 1860, at the beginning of the century there had been "large investments in salt-works." Freeman, *The History of Cape Cod*, vol. 2, 251.
69. Ibid., 696.
70. Ibid., 696.
71. Freeman, *The History of Cape Cod*, vol. 1, 757, 607.
72. Ibid., and Berger, *Cape Cod Pilot*, 98, 99.
73. This compares to the Falmouth Bank, which was incorporated the same year for $100,000. Freeman, *The History of Cape Cod*, vol. 1, 617.
74. Ibid., vol. 2, 540.
75. Deyo, *History of Barnstable County*, 125, 145.
76. Freeman, *The History of Cape Cod*, vol. 2, 354, 740. Early-nineteenth-century Orleans sent close to 500 bushels of corn to Boston and had more than one-third of its tillage land in grain. It also produced enough vegetables, meat, and butter to meet its own needs, 723.
77. Forest cover on the Outer Cape fell from over 70 percent in the early colonial period to less than 20 percent by the middle of the nineteenth century.
78. Friedman, "Governing the Land," 211.
79. Shallow kettle ponds filled with vegetable matter, particularly pine needles and oak leaves which are highly acidic. Due to the high accumulation of this organic matter the levels of dissolved oxygen declined, creating the context for the production of peat.
80. Jonathan Lincoln noted, "Peat was cheap and made a fairly good fire to cook by but . . . how it would smoke." Lincoln, *Yesterdays*, 117.
81. Dwight, *Travels*, vol. 3, 56.
82. Sarah Augusta Mayo, *Looking Back, 1830–1870*, ed. Janine M. Perry (Brewster, MA: Brewster Ladies Library Association, 2003), 111; Freeman, *The History of Cape Cod*, vol. 2, 724, 740. Freeman (740) noted that by the middle of the nineteenth century imported coal was increasingly substituted for local peat.

83. First the bog had to be cleared of bushes and stumps and the surface leveled then covered with 5 inches of sand. The vines had to be then set out and competing vines and weeds removed. By the middle of the nineteenth century it took roughly two to five hundred dollars an acre to establish a bog. Deyo, *History of Barnstable County,* 150.
84. Edwards, *Cape Cod New and Old,* 76.
85. *Historic Cultural Land Use Study of Lower Cape,* 27, 28. The completion of the Erie Canal in 1826 drove down the price of grain, particularly wheat. Cape farmers, whose wheat production was profitable only because of the poor wheat harvests of most of the rest of New England, gave up attempting to compete with western farmers in the production of grains.
86. In 1812 Samuel Wing got permission to build a new dam and mill, this time for cotton production. Deyo, *History of Barnstable County,* 275.
87. Friedman, "Governing the Land," 254.
88. Quoted in Freeman, *The History of Cape Cod,* vol. 2, 30.
89. Sheep raising grew dramatically on the Cape in the early nineteenth century and supplied the wool for the Cape's growing woolen mills.
90. Freeman, *The History of Cape Cod,* vol. 2, 465. Ruth Friedman puts the date of this event as 1804.
91. Quoted in Friedman, "Governing the Land," 201.
92. Increasingly in the nineteenth century mill owners managed to avoid the common-law requirement to allow the migration of fish to their spawning grounds. The courts were losing interest in these common laws and the General Court increasingly favored mill owners over common-law traditions. Although increasingly favoring the mill owners, the General Court did pass legislation in 1858 preventing seine or dragnetting within a mile and half of the mouth of any river or stream from March through September. Freeman, *The History of Cape Cod,* vol. 2, 484.
93. In 1802 the town of Falmouth accused Barnabus Hinckley of violating the 1798 law requiring him to open his dam for migrating fish, and he was found guilty. Hinckley appealed his case and later won his appeal. Under the leadership of Joseph Dimmick, a leader of the Herring Party, the town voted to appeal to the legislature for a statute protecting fish migration. But the Mill-dam Party rallied and created a committee to oppose Dimmick's appeal. Friedman, "Governing the Land," 196.
94. Thoreau, *Cape Cod,* 92.
95. Friedman, "Governing the Land," 271.
96. Freeman, *The History of Cape Cod,* vol. 1, 659.
97. See Friedman, "Governing the Land," 270–72, 293, and 337 for discussion of the increased role of private companies in attempts to control nature.
98. Henry F. Walling, "The 1858 Map of Cape Cod, Martha's Vineyard and Nantucket" (Cape Cod, MA: Cape Cod Publications, 2009), 14.
99. Berger, *Cape Cod Pilot,* 209, 211.
100. Quoted in Friedman, "Governing the Land," 218.
101. See Friedman (218, 219) for a discussion of this reluctance.
102. Dwight, *Travels,* vol. 3, 58.

103. Kendall did not understand Cape farming. His comparison was with English farming and English soil. He focused on the poor sandy soil and the use of horseshoe crabs as fertilizer. He also believed that the only way Cape Codders could survive was to harvest the bounty of the sea. He did not notice that Cape farmers used manure gathered over the winter from stock fed on marsh hay. His view of the Cape as an impoverished area, although not appropriate to the Cape of 1808, foreshadowed what would increasingly be the reality of Cape Cod agriculture. After publishing his travels in North America, Kendall went on to write children's stories that gave animals voices and personalities. Edward Augustus Kendall, *Travels Through the Northern Parts of the United States, in the Years 1807 and 1808* (New York: I. Riley, 1809), 148–65; Freeman, *The History of Cape Cod*, vol. 1, 721.
104. Thoreau, *Cape Cod*, 149.
105. Ibid., 157.
106. Ibid., 149. For Thoreau even the thin woods, mostly pitch pine, were inferior to the white pine and oak that had existed in the eighteenth century.
107. Ibid, 297.
108. Ibid., 237.
109. Ibid., 151.
110. Ibid., 25, 28.
111. David R. Foster and Glenn Motzkin, "Ecology and Conservation in the Cultural Landscape of New England: Lessons from Nature's History," *Northeastern Naturalist* 5 (1999): 111–26.
112. Deyo, *History of Barnstable County*, 470.
113. Freeman, *The History of Cape Cod*, vol. 2, 537.
114. Ibid., 538.
115. Thoreau, *Cape Cod*, 41.
116. Edwards, *Cape Cod New and Old*, 117, 118.
117. Thoreau, *Cape Cod*, 296.
118. Freeman, *The History of Cape Cod*, vol. 2, 723.
119. Ibid., 354.
120. Thoreau, *Cape Cod*, 159.
121. Echeverria, *A History of Billingsgate*, 38.
122. Freeman, *The History of Cape Cod*, vol. 2, 586.
123. Dwight, *Travels*, 61.
124. Legislative Committee of 1824, "Report of the Legislative Committee of 1824," (Boston, 1825).
125. Friedman, "Governing the Land," 325.
126. Freeman, *The History of Cape Cod*, vol. 2, 537.
127. Dwight, *Travels*, 56.
128. Thoreau, *Cape Cod*, 40.
129. Berger, *Cape Cod Pilot*, 89.
130. Even as shipbuilding declined in the 1850s, Dennis still employed more than fifty workers building boats in its yards. Deyo, *History of Barnstable County*, 142.

131. Deyo, *History of Barnstable County,* 142.
132. Smith and Oliver, "Port on the Bay."
133. Deyo, *History of Barnstable County,* 859, 929.
134. Thoreau, *Travels,* 151.
135. Freeman, *The History of Cape Cod,* vol. 2, 754. See also Deyo, *History of Barnstable County,* 515.
136. Smith and Oliver, "Port on the Bay."
137. Edwards, *Cape Cod New and Old,* 77.
138. Deyo, *History of Barnstable County,* 979.

5. THE DECLINE OF THE ESTABLISHED ECONOMY

1. Bureau of Statistics of Labor, *Census of Massachusetts, 1885,* vol. 1, "Population and Social Statistics," part 2, ed. Carroll Wright (Boston: Wright and Potter Printing Company, 1888), 34, 35, 823.
2. Quoted in Commonwealth of Massachusetts, Dept. of Labor and Industries, "Population and Resources of Cape Cod" (Boston: Wright and Potter Printing Company, 1922), 10.
3. 1816, the worse of these two hard years, was known as the year without a summer, when much of New England faced frost all summer long.
4. *Census of Massachusetts, 1885,* vol. 3, "Agriculture."
5. Frederick Freeman, *The History of Cape Cod: The Annals of Barnstable County and of Its Several Towns Including the District of Mashpee* (Boston: George Rand and Avery, 1862), vol. 2, 249. Despite the importance of the Great Meadow to Barnstable's agriculture, the state incorporated the "Great Marsh Diking, Water Power and Fishing Co." in 1850 "to reclaim for more useful purposes the vast amount of salt marshes that indent the bounds of the Cape." The company proposed to construct a dike from Calves pasture to Sandy neck, "to prevent the flow of salt waters . . . for the purpose of draining the marshes there situate and converting the same into meadow or tillage land." Freeman, *The History of Cape Cod,* vol. 1, 659.
6. There is some question as to the exact date of the formation of the Agricultural Society, 1843 or 1844. For our purposes it makes no difference so I picked the later year, when the society was definitely engaged in activity.
7. By 1850, although all of New England was significantly deforested, the Cape was far more deforested than the rest of the region. Only 20 percent of the Outer Cape, for example, was forested at that period, with less than 13 percent of Eastham in woods. See David R. Forester, "Land-Use History (1730–1990) and Vegetation Dynamics in Central New England, *Journal of Ecology,* 80 (1992): 753–72; David R. Forester and John F. O'Keefe, *New England Forests Through Time: Insights from the Harvard Forest Dioramas* (Cambridge: Harvard University Press, 2000). See also Gordon G. Whitney, *From Coastal Wilderness to Forested Plain: A History of Environmental Change in Temperate North America from 1500–the Present* (New York: Cambridge University Press, 1994); and L. Stanford Altpeter, "A History of Forests of Cape Cod" (MS thesis, Harvard University, 1939), 49.

8. With the war's end Dennis sent out 48 vessels with 722 men on board to the fishing grounds of the North Atlantic, representing over $117,000 in capital. Another 84 boats with 445 men worked coastal fishing. Agnes Edwards, *Cape Cod New and Old* (Boston: Houghton Mifflin, 1918), 74. What Dennis workers did not return to were the 85 old saltworks on its shore (Edwards, 74). Few also went back to fishing. Twenty years after the war, Dennis sent not 132 but only 60 vessels out to catch fish either, on the banks or near-shore. Simeon Deyo, *History of Barnstable County, Massachusetts* (New York: H. W. Blake, 1890), 516.
9. Orleans, which sent out four whaling ships employing 125 men in 1855, bringing in $20,000 in whale oil a year, sent out no whaling boats in the postwar period. Edwards, *Cape Cod New and Old*, 105.
10. *Census of Massachusetts, 1885*, vol. 2, "Manufacturing, Fisheries, and Commerce." Cheap petroleum-based kerosene lowered the demand for whale oil, but did not end it. Demand continued to be high into the 1870s and encouraged whalers to move into the rough frigid waters of the northern Pacific. The problem for Cape whalers was cost. The costs of whaling operations increased dramatically and quickly outpaced the capacity of most Cape whalers, while Nantucket, New Bedford, and Honolulu continued to search for ever-scarcer whales and new whaling grounds. The price of whale oil did begin to drop in the 1870s with the acceptance of kerosene, and even the Nantucket and New Bedford whalers were having difficulty surviving, but by this time, for the most part, Cape Codders were out of the business of killing whales. See Peter Nichols, *Oil and Ice: A Story of Arctic Disaster and the Rise and Fall of America's Last Whaling Dynasty* (New York: Penguin Press, 2009).
11. Deyo, *History of Barnstable County*, 666–70.
12. By 1885 Provincetown's 36 whaler men and Mashpee's seven were the only ones left out of the hundreds that earlier had left the ports of various Cape towns. *Census of Massachusetts, 1885*, vol. 1, "Population and Social Statistics," part 2, 50.
13. Dos Passos and Shay argued in 1947 that it was the increasing time at sea, four to five years, and the declining returns for the smaller capitalized Cape operations that ended whaling for Cape Codders. Katharine Dos Passos and Edith Foley Shay, *Down Cape Cod* (New York: Robert McBride, 1947), 13. See also Commonwealth of Massachusetts, Dept. of Labor and Industries, "Population and Resources of Cape Cod" (Boston: Wright and Potter, 1922), 10.
14. Freeman noted that although cod was available in large numbers in local waters in the eighteenth century, by the nineteenth they could only be found "on other localities distant," particularly the banks off of Newfoundland and the Strait of Belle Isle off of Labrador. Freeman, *The History of Cape Cod*, vol. 2, 645. For cod, although the catch varied from year to year, the shift from small sloops (with six to eight workers with lines over the side) to large schooners (with 25 men setting out dories with long lines and baited hooks) doubled the catch between the first half of the century and 1880, but overfishing began to take its toll. By the late 1880s the catch had fallen back to its first-half amount. Deyo, *History of Barnstable County*, 132, 133.
15. Freeman, *The History of Cape Cod*, vol. 2, 622.

16. With the arrival of the railroad, Provincetown and Chatham were sending fresh fish on ice to Boston and New York markets. Provincetown had a fish freezing plant in 1893 and even Truro, despite a significant decline in its fisheries, still had two canning factories at the end of the century. "Historic and Cultural Land Use Study of Lower Cape Cod," manuscript (Wellfleet, MA: National Seashore Library), 28.
17. Beginning in 1791 the right to fish for herring to sell, as opposed to eat, was auctioned off by the town selectmen. Town citizens were allowed to take 200 for personal consumption. The proceeds from the town auctions of the fishing rights were enough to pay all the town officials. Town of Wellfleet, *Wellfleet Annual Reports* 1880–1905.
18. Edwards, *Cape Cod New and Old*, 134.
19. In 1885 Provincetown harvested 16 million pounds of cod. Although fishing declined across much of the Cape in the second half of the nineteenth century, with the annual catch declining 40 percent from 1885 to 1895, Provincetown and Chatham continued to catch and market significant amounts of fish.
20. Freeman, *The History of Cape Cod*, vol. 2, 178.
21. Between 1845 and 1865 Wellfleet had, besides its boats chasing cod, over 100 vessels fishing for mackerel.
22. Besides bluefish and mackerel, large numbers of flounder, alewives, shad, eels, and striped bass were taken in the spring. In the summer bluefish and mackerel filled the pound nets. Frederick True, "Pound-Net Fisheries of the Atlantic States" in *Fisheries and the Fishery Industries of the United States,* ed. George Brown Goode (Washington DC: Government Printing Office, 1887), section 5, vol. 1, 595–610.
23. Henry C. Kittredge, *Cape Cod: Its People and Their History* (Boston: Houghton Mifflin, 1930), 199.
24. See Matthew McKenzie, *Clearing the Coastline: The Nineteenth-Century Ecological and Cultural Transformation of Cape Cod* (Hanover, NH: University Press of New England, 2010) for an excellent discussion of the conflict over pound netting. McKenzie's discussion of the conflict between the pound netters and the non-pound-netting fishermen and the role of the scientific experts and politics in this conflict is one of the best discussions one will ever find of this kind of issue and is a must-read for anyone interested in U.S. fisheries.
25. Kittredge, *Cape Cod,* 194.
26. "Population and Resources of Cape Cod," 10. Dennis and Harwich also switched to mackerel because the silting in of their harbors prevented their cod boats from getting across the sand bars.
27. "Population and Resources of Cape Cod," 11.
28. Wellfleet's mackerel fleet grew in the first two decades after midcentury. In 1870 the town had 100 schooners chasing mackerel and more than 1,500 fishermen were fishing for mackerel.
29. Deyo, *History of Barnstable County,* 791.
30. Horace Greely Wadlin, "Social and Industrial Changes in the County of Barnstable," *Massachusetts Bureau of Statistics of Labor* (Boston: 1897), 51, 52.
31. Ibid., 62.

NOTES TO PAGES 87–91   237

32. Ibid., 52.
33. Ibid., 79.
34. Deyo, *History of Barnstable County*, 791, 792.
35. Ibid., 142.
36. A 50–75 ton schooner could take four or five boatwrights up to four months to build.
37. Wadlin, "Social and Industrial Changes," 79.
38. The gale of 1841 destroyed a large number of Cape vessels, along with many of the Cape's ship works. John Braginton Smith and Duncan Oliver, "Port on the Bay, Yarmouth's Maritime History on the North Sea, 1638–the Present" (Yarmouth, MA: Historical Society of Old Yarmouth), Cape Cod Community College Library, Nickerson Rm. 36.
39. Arthur W. Tarbell, *I Retired to Cape Cod* (New York: Stephen Daye, 1944), 46. A single run of a clipper ship from Boston to China could pay back the cost of building and rigging a clipper ship.
40. *Census of Massachusetts, 1885*, vol. 1, "Population and Social Statistics," part 2, 42, 50.
41. Everett I. Nye, *History of Wellfleet from Early Days to Present Time* (Hyannis, MA: F. B. & F. P. Goss, 1920).
42. Several factors hurt the salt industry. The reduction of the tariff on foreign salt, mostly from Portugal, increased the supply of salt and cut its price, as did competition from Syracuse, New York, salt producers. The practice of freezing fish, although not ending the demand for salt by the fishing industry, did reduce it. The decline of locally available wood, often free to local salt producers who owned nearby woodlots, and the increased cost of importing lumber from Maine, put the nail in the coffin of the Cape's salt industry.
43. The decline in saltworks occurred across the Cape. The elimination of the federal bounty on salt and competition from foreign sources (after the tariff was lowered) as well as western mines did hurt salt manufacturers, but high-quality sea salt was still in demand, especially among fishermen. Ultimately it was the increased cost of pine, no longer locally available for making or repairing the vats, that "was a check upon the business." Deyo, *History of Barnstable County*, 125, 145.
44. Katherine Crosby, *Blue-Water Men and Other Cape Codders* (New York: Macmillan, 1946), 255.
45. Agents of the railroad fanned out over the Cape selling shares. By the end of the century owning a few shares of the Old Colony Rail Road was considered a sign of being an old-timer.
46. Even into the late 1850s Thoreau complained of the quality of the railroad service on the Cape.
47. Deyo, *History of Barnstable County*, 119.
48. Monomoy Point had a thriving harbor with warehouses, wharves, and a significant fishing fleet in 1850, but by the twentieth century its buildings were empty and falling down, its wharves were gone, and sand filled the harbor. Joseph Crosby Lincoln, *Cape Cod Yesterdays* (Boston: Little, Brown, 1935), 161.
49. It was not just the railroad that hurt the packet service. Packet service to Truro ended

even before the arrival of the railroad because of the silting in of its harbor. Deyo, *History of Barnstable County,* 120, 121.
50. Edward Rowe Snow, *A Pilgrim Returns to Cape Cod* (Boston: Yankee Publishing Company, 1945), 202.
51. Freeman, *The History of Cape Cod,* vol. 1, 663; Snow, *A Pilgrim Returns,* 204.
52. Katherine Lee Bates, "Memoir," quoted in Scott Corbett, *Cape Cod's Way: An Informal History of Cape Cod* (New York: Thomas Crowell, 1955), 56.
53. Lincoln, *Cape Cod Yesterdays,* 83.
54. Lincoln estimated that thousands of Cape Codders responded to the solicitation of the Old Colony Rail Road to buy stock so the line could be extended to Provincetown. The Old Colony Rail Road held "stockholders day," once a year for the yearly meeting of the company. On stockholders day a stockholder rode free from the Cape to Boston and back. Lincoln, *Cape Cod Yesterdays,* 27. The Old Colony Rail Road also encouraged the town-funded extensions into the towns not serviced by the main line. Chatham did not get rail service until 1887, when the Chatham Rail Road Company built its extension into the town. The largest stockholder of the Chatham Rail Road Company was the town of Chatham. Joshua Atkins Nickerson II, *Days to Remember: A Chatham Native Recalls Life of Cape Since the Turn of the Century* (Chatham, MA: Chatham Historical Society, 1988), 56.
55. Salt marshes transfer food and chemicals between land and water. They are highly productive biological communities found in the intertidal areas up tidal rivers or behind barrier beaches. One, the Barnstable Marsh, was two miles wide and four miles long. Each year between 5 and 10 tons of organic matter are produced by each acre of salt marsh. This compares to 4 tons per acre of a hay field and 1.5 tons of a wheat field. Over time, decaying marsh plants accumulate under the living mat of vegetation. High acidity and lack of oxygen encourage peat formation. Across the surface of the tidal marsh single-cell algae bloom and form the basic food stuff for insects and small crustaceans. Blue-green algae are nitrogen fixers, pumping nutrients into the system, which other plants take up. The building of the railroad beds disrupted and often destroyed this productive ecosystem. Carl Carlozzi et al., "Ecosystems and Resources of the Massachusetts Coast," Massachusetts Coastal Zone Management, Institute for Man and Environment, 12–15. See also John Portnoy and Michael Soukup, "From Salt Marsh to Forest: The Outer Cape Wetlands," *The Cape Naturalist,* 11 (Fall 1982): 29–34.
56. The strain of phragmites that moved into these wetlands was originally the North American strain. Today both the North American phragmites and a foreign strain from Europe are invasive. The European *Phragmites australis* began to take over American wetlands in the late nineteenth century.
57. Wadlin, "Social and Industrial Changes," 62.
58. Freeman, *The History of Cape Cod,* vol. 1, 627. Initially the glass-works had one eight-foot furnace with each pot holding 800 pounds and producing 7,000 pounds of glass with 60 workers. By midcentury its capital had expanded to well over 400,000 and it was producing 100,000 pounds of melt weekly, worth over $600,000.
59. Deyo, *History of Barnstable County,* 915.

60. The land Perry developed was primarily empty, abandoned farmland. The 1858 Map of Cape Cod, Martha's Vineyard, and Nantucket, by Henry F. Wallings (Cape Cod: Cape Cod Publications, 2009), Map 30, shows only three houses on the area Perry would develop and none of those were near the shore. Two of them were in the Perry family.
61. Perry, the son of Captain Caleb Perry, was raised in an old Cape house on Country Road, Monument Beach, in Bourne. In 1900 Perry developed a large pasture bordering Phinney's Harbor. He laid out streets and lots and began selling off the land. This enterprise was so successful that he opened an office in Monument Beach and continued developing and selling land along Buzzards Bay. E. G. Perry, *A Trip Around Cape Cod: Our Summer Land and Memories of My Childhood* (Boston: Charles Binner Publisher, 1895), 399.
62. In his book Perry asked, "Do you want to invest some money that will bring you big returns? There has been a great deal of money made in buying land on Cape Cod for the last 10 years and in the next few years to come there is still a greater chance to make money for the land is increasing in valuation very rapidly each year. I have some extra fine tracts of land for sale where in a few years money invested can be more than doubled. I have some very reasonable estates and some of the handsomest estates on Cape Cod for sale. I have them in nearly every location on the cape." Perry, *A Trip Around Cape Cod*, 399.
63. Ibid.
64. Edwards, *Cape Cod New and Old*, 63.
65. In 1881 Wellfleet sent out 31 boats fishing for mackerel. They returned with 35,677 barrels of fish. Five years later 29 boats were only able to bring in 3,566 barrels. The catch went up and down over the next few years but the trend was downward. Thirty boats caught 4,832 barrels in 1888; the next year only 21 boats went out, and they caught only 1,697 barrels. Other Cape towns had similar declines. Deyo, *History of Barnstable County*, 137.
66. Henry David Thoreau, *Cape Cod* (New York: Penguin, 1987; orig. 1865), 318.
67. More typical for the wealthy at midcentury was to vacation in rural Vermont, Newport, Rhode Island, or, as in the case of Henry Ingersoll Bowditch, on the Isles of Shoals. Henry Ingersoll Bowditch letter to Olivia Bowditch, August 18, 1858, Henry Ingersoll Bowditch Papers, Countway Library, Harvard University. For a discussion of the recreational habits of the well-to-do at midcentury see Dona Brown's excellent study of the development of recreation in the nineteenth century: Dona Brown, *Inventing New England: Regional Tourism in the Nineteenth Century* (Washington, DC: Smithsonian Institution Press, 1995).
68. Thoreau, *Cape Cod*, 74.
69. As early as 1837 the *Yarmouth Register* noted, "The sports season is at hand; and our Boston friends will soon seek relaxation from the cares and toils of business and come to sport on marshes and fish in our ponds and streams." Ruth Lynn Friedman, "Governing the Land: An Environmental History of Cape Cod, Massachusetts, 1600–1861" (PhD thesis, Brandeis University, 1992), 202.
70. See John T. Cumbler, *Reasonable Use: The People, the Environment and the State, New*

England 1790–1930 (New York: Oxford University Press, 2001), 171–76, for a discussion of sports hunting and fishing and manliness.

71. Dennis and Marion Chatham, *Cape Coddities* (Boston: Houghton Mifflin, 1920), 102, 103; Deyo, *History of Barnstable County,* 137.
72. Hunting game birds was not just the activity of the wealthy sportsmen. Until 1918 the Cape had a thriving industry in market hunting of ground birds, ducks, and geese. Hundreds of these birds were killed by "market gunners" who sent their kill to the markets of Boston and New York, where they entered the stomachs of well-to-do purchasers through both home-cooked and restaurant meals. After the 1918 treaty with Canada outlawed the selling of game birds, market gunning disappeared, while duck, geese, and ground-bird hunting became solely a seasonal sport. Walter Teller, *Cape Cod and Off Shore Islands* (Englewood Cliffs, NJ: Prentice Hall, 1970), 177.
73. Daniel Webster was one of many famous New Englanders who traveled down to the Cape to hunt and fish. Theodore Lyman, from one of Boston's most elite families, regularly took the train down to the Upper Cape to fish. See Theodore Lyman III, Diaries, April, 1, 8, and 28, May 18 and 26, 1874, vol. 35, Lyman Family Papers, Massachusetts Historical Society.
74. The Chathams, in *Cape Coddities* (102–103), noted that "a number of gentlemen . . . have built small camps upon certain of these secluded spots for casual excursions." He added, "By leaving Boston at noon they can be in camp by sundown." Typical of these hunting and fishing clubs were the Brant Club on Monomoy Beach and the Hunters Hotel in Chatham. Lincoln, *Cape Cod Yesterdays,* 133.
75. Loring Underwood continued to hunt while on the Cape, but was unable to interest his daughters (he had no sons) in sport shooting. The family did enjoy entertaining a wide range of influential and artistic friends and relatives. Underwood was not only a founding member of a local hunting club, but also was influential in founding the Eastward Ho Country Club, whose grounds he designed. Conversations with Loring Underwood's great-granddaughter Pamela Mack; Carol Shloss, "Preserving the Light: The Photography of Wm. Lyman Underwood and Loring Underwood," in *Gentlemen Photographers: The Work of Loring Underwood and Wm. Lyman Underwood,* ed. Robert Lyons (Boston: Solio Foundation, Northeastern University Press, 1989).
76. Thoreau, *Cape Cod,* 315. See Brown, *Inventing New England* for a discussion of the role of camp meetings in pioneering beach vacations for the middle classes.
77. Freeman, *The History of Cape Cod,* vol. 1, 639; Edward Rowe Snow, *A Pilgrim Returns to Cape Cod,* 153.
78. Edwards, *Cape Cod New and Old,* 63. Joseph Lincoln noted that the Old Colony Railroad ran special trains to the campgrounds. Lincoln, *Cape Cod Yesterdays,* 24.
79. Thoreau, *Cape Cod,* 76.
80. Edwards, *Cape Cod New and Old,* 63.
81. James C. O'Connell, *Becoming Cape Cod: Creating a Seaside Resort* (Hanover, NH: University Press of New England, 2003), 5.
82. "Population and Resources of Cape Cod," 78.
83. Florence Baker, *Yesterday's Tides* (Yarmouth, MA: self-published, 1941), 197, 211, 215.

84. Ibid., 207, 208.
85. For more on relatives vacationing on the Cape see Baker, *Yesterday's Tides*, 205–22, 275–85.
86. Deyo, *History of Barnstable County*, 671.
87. Ibid., 411.
88. Altpeter, "A History of Forests," 66.
89. Deyo, *History of Barnstable County*, 156.
90. Ibid., 417.
91. "Population and Resources of Cape Cod," 50.
92. Deyo, *History of Barnstable County*, 156.
93. Wadlin, "Social and Industrial Changes."
94. Quoted in Arthur W. Tarbell, *I Retired to Cape Cod* (New York: Stephen Daye, 1944), 95.
95. Marcia Monbleau, *Pleasant Bay: Stories from a Cape Cod Place* (Harwich, MA: Friends of Pleasant Bay, 1999), 13.
96. Ibid., 14.
97. Deyo, *History of Barnstable County*, 415.
98. Tarbell, *I Retired to Cape Cod*, 96.
99. Millard Faught, *Falmouth, Massachusetts: Problems of a Resort Community* (New York: Columbia University Press, 1945), 36, 51.
100. Ibid., 51. See also Josef Berger (Jeremiah Digges), *Cape Cod Pilot* (Cambridge, MA: MIT Press, 1969; orig. 1937), 369.
101. Deyo, *History of Barnstable County*, 639; Faught, *Falmouth*, 36.
102. Faught, *Falmouth*, 36.
103. Deyo, *History of Barnstable County*, 640.
104. E. G. Perry, *A Trip Around Buzzards Bay Shores* (Bourne, MA: Bourne Historical Commission, 1976), 117.
105. Ibid., 150.
106. Millard Faught argued that it was outsiders who profited by the increase in land value with the building of summer estates—not locals, who received "very low prices" for their land. He also noted that the estates did not displace older land uses, but moved into areas where older uses were being abandoned. Faught, *Falmouth*, 52–53.
107. Faught, *Falmouth*, 53 (my italics for emphasis).
108. Ibid., 36.
109. Perry, *A Trip Around Buzzards Bay Shores*, 80, 87.
110. Ibid., 87.
111. Deyo, *History of Barnstable County*, 407.
112. Ibid., 795.

6. TRAINS, CARS, COTTAGES, AND RESTAURANTS

1. Commonwealth of Massachusetts, Dept. of Labor and Industries, "Population and Resources of Cape Cod" (Boston: Wright and Potter Printing, 1922), 30.
2. Anthony Marshall's family farm in Truro was typical. In the fall they planted rye on

their fields to restore nitrogen to the soil and protect it from being blown away. Over the winter the family mixed cordgrass drift with barnyard manure. Cordgrass and seaweed would also be packed in around the home's foundation to a depth of about two feet for insulation during the winter. In the spring this would be mixed with cordgrass and manure in the barnyard, making compost that was spread on the fields. Anthony L. Marshall, *Truro, Cape Cod as I Knew It* (New York: Viking Press, 1974), 147, 157.

3. Elroy Thompson noted that by the mid-1920s "strawberry raising, poultry raising and agriculture in general had made remarkable strides." However, he also believed that Cape agriculture was being held back because of lack of transportation to "the markets of the world." Elroy S. Thompson, *History of Plymouth, Norfolk and Barnstable Counties*, vol. 2 (New York: Lewis Publ., 1928), 881.

4. Horace Greeley Wadlin, "Social and Industrial Changes in the County of Barnstable" (Boston: Massachusetts Bureau of Statistics of Labor, 1897), 102.

5. The asparagus crop turned Eastham into "asparagus country," but a rust fungus destroyed the crop in the years following World War I. In the 1920s a new rust-resistant strain led to a revival of asparagus, but it never regained its turn-of-the-century importance. Josef Berger (Jeremiah Digges), *Cape Cod Pilot* (Cambridge, MA: MIT Press, 1969; orig. 1937), 182.

6. Wadlin, "Social and Industrial Changes," 56. Katharine Dos Passos and Edith Foley Shay, *Down Cape Cod* (New York: Robert M. McBride, 1947), 25; Whitman and Howard, *What the Herring River Dike Project Would Mean To The Town of Wellfleet: Report of Messrs. Whitman and Howard on Proposed Dike At Herring River* (Wellfleet, MA, 1906), 12.

7. E. G. Perry, *A Trip around Cape Cod: Our Summer Land and Memories of My Childhood* (Boston: Charles Binner Publisher, 1898), 18.

8. Dos Passos and Shay, *Down Cape Cod,* 25. Katharine Dos Passos was a lover of Cape Cod. She and her husband, the author John Dos Passos, were central figures in the growing intellectual community of Provincetown in the post–World War I period.

9. "Population and Resources of Cape Cod," 30.

10. Wadlin, "Social and Industrial Changes"; Simeon Deyo, *History of Barnstable County, Massachusetts* (New York: H. W. Blake, 1890), 8; Thompson, *History of Plymouth, Norfolk and Barnstable Counties,* vol. 2, 487, 849.

11. "Population and Resources of Cape Cod," 34. By 1920 these mostly Portuguese farms were small affairs of one or two acres in strawberries producing about 300 bushels per acre or $1,200 an acre of strawberries. Half the strawberries went to Cape consumers while the rest, some half-million quarts, were shipped to cities on the East Coast. In 1915 the strawberry growers formed the Cape Cod Strawberry Growing Association to protect their interests. "Population and Resources of Cape Cod," 34; Dos Passos and Shay, *Down Cape Cod,* 24, 25, 27; Thompson, *History of Plymouth, Norfolk and Barnstable Counties,* vol. 2, 849.

12. Dos Passos and Shay, *Down Cape Cod,* 25.

13. Wadlin, "Social and Industrial Changes," 55.

14. Lewis M. Alexander, "The Impact of Tourism on the Economy of Cape Cod, Massa-

chusetts," *Economic Geography* 29 (October. 1953): 322. In 1890 the Cape still had 29 percent of its land in agriculture. Some of that was for dairy operations (there were more than 1,500 dairy cows on the Cape as late at 1920) and some of it was for pasture for horses used in transportation and as draft animals (324).

15. Thompson, *History of Plymouth, Norfolk and Barnstable Counties,* vol. 2, 487. Thompson (829) noted that by the 1920s roadside farm stands had become a common sight on Cape Cod, furnishing "an outlet for thousands of dollars worth of fruits, vegetables, honey, poultry, berries . . . and every other conceivable thing in the name of souvenir and novelty all through the marketing season."
16. Thompson, *History of Plymouth, Norfolk and Barnstable Counties,* vol. 2, 871.
17. Ibid., 852; Alexander, "The Impact of Tourism," 324.
18. Not all Cape farms were small affairs. In 1920 the Coonamessett Ranch held some 14,000 acres on the corner of Bourne, Sandwich, Falmouth, and Mashpee. Most of the land was in forests, but the ranch had 500 acres in crop production, a huge market garden for vegetables, and more than 100 cows producing milk to sell to summer residents. The ranch was to be a demonstration of the viability of agriculture on the Cape but it did not survive the Depression. "Population and Resources of Cape Cod," 38.
19. By 1940 Barnstable County realized $2,000,000 in agricultural income. Millard C. Faught, *Falmouth, Massachusetts: Problems of a Resort Community* (New York: Columbia University Press, 1945), 169. Falmouth farmers, for example, sold 160,587 gallons of milk and 400 pounds of cheese mostly to local visitors, hotels, and guest homes in 1939. Falmouth, which still had 1,500 acres of land in agricultural production had 26 sheep, 2,743 chickens, 61 horses, 190 hogs, and 379 cattle and produced over 18,300 bushels of potatoes. Faught, *Falmouth,* 168.
20. Marshall's family sold their produce to roadside stands and to Provincetown. Marshall, *Truro,* 143, 156, 157. The Marshall farm included several separate parcels of land. Besides the family homestead they had a nine-acre woodlot as well as a hay field in Wellfleet.
21. Wadlin, "Social and Industrial Changes," 59.
22. Ibid., 57. See Matthew McKenzie, *Clearing the Coastline: The Nineteenth Century Ecological and Cultural Transformation of Cape Cod* (Hanover, NH: University Press of New England, 2010) for an excellent discussion of the conflicts between the hook and line fishermen and the pound netters.
23. Wadlin, "Social and Industrial Changes," 57; Dos Passos and Shay, *Down Cape Cod,* 99.
24. Wadlin, "Social and Industrial Changes," 57. A half-century earlier Wellfleet's fishing industry was second in the state just behind Gloucester's, but by the twentieth century it had fallen far behind (39).
25. E. C. Janes, *I Remember Cape Cod* (Brattleboro, VT: Stephen Greene Press, 1974), 49.
26. Dos Passos and Shay, *Down Cape Cod,* 88.
27. Richard Holmes et al., "Historic Cultural Land Use Study of Lower Cape Cod: A Study of the Historical Archeology and History of the Cape Cod National Seashore and the Surrounding Region," National Park Service, U.S. Department of the Interior Lowell, MA: University of Massachusetts, Archeological Branch: 1968), 30. In 1911 the

state of Massachusetts passed legislation allowing the Cape towns to set aside grants for propagating clams and oysters. The development of these oyster grants led to the revival of Wellfleet's oysters and its oyster industry. Thompson, *History of Plymouth, Norfolk and Barnstable Counties*, vol. 2, 830.

28. Dennis and Marion Chatham, *Cape Coddities* (Boston: Houghton Mifflin, 1920), 136–39.
29. Katharine Dos Passos and Edith Shay noted that the Crosby boatyard in Osterville was adapting his old fishing catboat for sport fishing and by the 1930s the boatyard was building "racing knockabouts" for the wealthy summer visitors to the Cape. Dos Passos and Shay, *Down Cape Cod*, 195. See also Wadlin, "Social and Industrial Changes," 53, and Alexander, "The Impact of Tourism," 325.
30. Janes, *I Remember Cape Cod*, 37.
31. See Dona Brown, *Inventing New England: Tourism in the Nineteenth Century* (Washington, DC: Smithsonian Institution Press, 1995) and Blake Harrison, *The View from Vermont: Tourism and the Making of an American Rural Landscape* (Burlington, VT: University of Vermont Press, 2006) for a discussion of the emergence of the appeal of the rural retreat for the wealthy.
32. Quoted in Brown, *Inventing New England*, 202.
33. Quoted in James C. O'Connell, *Becoming Cape Cod: Creating a Seaside Resort* (Hanover, NH: University Press of New England, 2003), 9.
34. Deyo, *History of Barnstable County*, 154. See 153 for discussion of tourism in Falmouth, but at the same time the author noted that on the Cape as a whole there was a shortage of good accommodations and that this should be remedied in the future.
35. Charles E. Fisher, *The Story of the Old Colony Railroad* (NP, 1919), 114.
36. Faught, *Falmouth* 51.
37. Ibid., 53.
38. See George Howe Colt, *The Big House: A Century in the Life of an American Summer Home* (New York: Scribner, 2003) for a description of the mosquitoes, brambles, and smells that greeted the first generation of wealthy visitors to the Cape and how these wealthy took advantage of cheap land and brought along their wealthy friends to establish colonies of summer estates.
39. Fisher, *The Story of the Old Colony Railroad*, 114.
40. Perry, *A Trip around Cape Cod*, 135.
41. Wadlin, "Social and Industrial Changes," 53. Wadlin noted the decline in fishing across the Cape. He commented there were many empty houses in Truro, and that in Wellfleet "weir fishing is done but at a loss, there are only 2 mackerel fishermen left out of a fleet of 63 which existed 10 years ago" (53). See McKenzie, *Clearing the Coast* for a discussion of the conflict between pound netters and sport fishermen.
42. Wadlin, "Social and Industrial Changes," 52, 53.
43. E. G. Perry, *A Trip around Buzzards Bay Shores* (Bourne, MA: Bourne Historical Society, 1976).
44. See Herrick and Newman, *Old Hyannis Port* (Hyannis, MA, 1968) for a discussion of the wealthy establishing summer estates in Hyannis Port at the turn of the century.
45. Perry, *A Trip around Cape Cod*, 218.

46. Ibid., 114, 104.
47. Ibid., 19, 26, 46, 57.
48. Ibid., 69, 57, 69.
49. Perry, *A Trip around Buzzards Bay Shores*, 284.
50. Perry, *A Trip around Cape Cod*, 104.
51. Perry, *A Trip around Buzzards Bay Shores*, 312, 313.
52. Perry, *A Trip around Cape Cod*, 9.
53. Ibid., 99.
54. Perry ended his book with a sales pitch: "I have some of the best tracts of land for sale to be found on the Bay," he wrote, "also a great many fine summer houses for sale and to rent." And: "I make a specialty of seashore property." Perry, *A Trip around Buzzards Bay Shores*, 150.
55. Joshua Nickerson retells the story of the demise of the Chatham Hotel as not a failure of vision but rather reflecting the reality that "tourists were not yet Outer Cape bound." Joshua Atkins Nickerson II, *Days to Remember: A Chatham Native Recalls Life on Cape Cod Since the Turn of the Century* (Chatham: Chatham Historical Society, 1988), 6.
56. Although popular with Chatham's wealthy summer residents, the country club had difficulty. When the Chatham Yacht Club, which shared the country club building, moved out, the club faced bankruptcy. In 1928 Roy Tomlinson, the president of "NaBisCo" ultimately rescued the country club, renaming it the Eastward Ho Country Club. Nickerson, *Days to Remember*, 18.
57. Nickerson, *Days to Remember*, 9.
58. Ibid., 9.
59. Janes, *I Remember Cape Cod*, 38.
60. Florence Baker, *Yesterday's Tide* (Yarmouth, MA: self-published, 1941), 288–90.
61. Janes, *I Remember Cape Cod*, 26.
62. Ibid.
63. Ibid., 26, 35.
64. Other resort areas also offered special attractions such as racetracks, boardwalks, or yacht racing, while sandy beaches were available all the way from New Jersey to Maine.
65. For Bostonians the trip by train to a resort hotel or to their summer estates was a manageable one-day affair, but for New Yorkers it was a far more complicated process including an overnight from New York to Fall River, then a train to Middleboro, and then a transfer to the train to Hyannis. Eleanor Early, *Cape Cod Summer* (Boston: Houghton Mifflin, 1949).
66. See Colt, *The Big House*.
67. Dos Passos and Shay, *Down Cape Cod*, 27.
68. Dos Passos and Shay, *Down Cape Cod*, 29.
69. Thompson, *History of Plymouth, Norfolk and Barnstable Counties*, vol. 2, 881.
70. Dos Passos and Shay, *Down Cape Cod*, 193.
71. Quoted in James Nickerson, "The Association for the Preservation of Cape Cod and the Natural Resources to Be Preserved" (Orleans, MA: APCC, 1977), 1.

72. Ibid., 2, 1.
73. Ibid., 2.
74. Perry, *A Trip around Cape Cod,* 9; Nickerson, "The Association," 1.
75. Baker, *Yesterday's Tide,* 285.
76. Massachusetts Highway Commission, *Fifth Annual Report,* January 1898, public document #54, 18.
77. Ibid. The Highway Commission allowed that pleasure travel should not be the only reason for building highways, but stated that the tourist industry would be "of such economic importance" to the commonwealth's future growth "that it should always be kept in mind." Massachusetts Highway Commission, *Fifth Annual Report,* 18.
78. Ibid., 23.
79. As early as 1898 Perry noted that the town of Bourne had engaged in a major road building and improvement campaign funded in large part by property taxes on the "summer friends," who contributed over 60 percent of the town taxes. Perry, *A Trip around Cape Cod,* 10.
80. "Population and Resources of Cape Cod," 61.
81. Janes, *I Remember Cape Cod,* 24–25, 19–20.
82. Thompson, *History of Plymouth, Norfolk and Barnstable Counties,* vol. 2, 105.
83. Agnes Edwards, *Cape Cod, New and Old* (Boston: Houghton Mifflin, 1918), xiii.
84. Ibid.
85. Chatham and Chatham, *Cape Coddities,* 71–85. In this work the Chathams talk about the importance of having a car on the Cape to get around to the inland ponds as well as to the small shops, villages, and small farmers selling antiques out of their barns or sheds.
86. See O'Connell, *Becoming Cape Cod,* 55, for a discussion of the impact of the car and mobility on the Cape's tourist industry.
87. Chatham and Chatham, *Cape Coddities,* 79–86; Thompson, *History of Plymouth, Norfolk and Barnstable Counties,* vol. 2, 829.
88. Deyo, *History of Barnstable County,* 115.
89. Edwards, *Cape Cod New and Old,* 206.
90. Faught, *Falmouth,* 130.
91. Dos Passos and Shay, *Down Cape Cod,* 197.
92. Ibid., 197. The United States Bureau of Fisheries established a research center in Woods Hole in 1871 and both the Oceanographic Institute and the Marine Biological Institute (established in 1888) called Woods Hole home.
93. Dos Passos and Shay, *Down Cape Cod,* 194.
94. Berger, *Cape Cod Pilot,* 379.
95. "Population and Resources of Cape Cod," 59–60.
96. Marcia Monbleau, *Pleasant Bay: Stories from a Cape Cod Place* (Harwich: Friends of Pleasant Bay, 1999), 30.
97. Ibid., 30. The camps drew a third of their campers from New York, a third from Boston, and a third from around the country and abroad.
98. Early, *Cape Cod Summer.*
99. Faught, *Falmouth,* 122.

100. Janes, *I Remember Cape Cod*, 207.
101. Ibid., 31–32.
102. Ibid., 87–99.
103. Ibid., 49.
104. In 1885 Falmouth reported to the state census taker that 11 of its male residents were gardeners. Falmouth was the only town to list gardeners as an occupation in 1885. The town also reported 193 of its female residents were employed in housework and another 74 females were servants.
105. Walter Teller, *Cape Cod and the Off Shore Islands* (Englewood Cliffs, NJ: Prentice Hall, 1970), 40.
106. Of the families visiting Falmouth in 1940, over half brought one or more live-in servants with them. A majority of the rest hired local help on either a regular or occasional basis. Faught, *Falmouth*, 46.
107. Chatham and Chatham, *Cape Coddities*, 17–25; L. Stanford Altpeter, "A History of the Forests of Cape Cod" (MS thesis, Harvard University, 1939), 66.
108. Faught, *Falmouth*, 108, 109.
109. Ibid., 109.
110. Katharine Crosby, *Blue-Water Men and Other Cape Codders* (New York: Macmillan, 1946), 126.
111. Nickerson, *Days to Remember*, 149.
112. Arthur W. Tarbell, *I Retired to Cape Cod* (New York: Stephan Daye, 1944), 31–32.
113. Altpeter, "A History of Forests," 67.
114. Alexander, "The Impact of Tourism," 324.
115. Ibid., 325.
116. Altpeter, "A History of Forests," 67.
117. Faught, *Falmouth*, 92.
118. Ibid., 3; Altpeter, "A History of Forests," 67.
119. Alexander, "The Impact of Tourism," 325,
120. Tarbell, *I Retired to Cape Cod*, 103.
121. Alexander, "The Impact of Tourism," 325.
122. Ibid., 324. Between 1920 and 1950 the Cape's cranberry acreage dropped by 1,000 acres.
123. Ibid., 323.
124. Hodgson Homes of Dover, Massachusetts, developed the first prefabricated homes in the United States. Ernest Hodgson began making prefabricated chicken coops in the 1890s and expanded to summer cottages in 1900. By midcentury the Hodgson Home became a popular middle-class means of quickly assembling an inexpensive summer cottage.
125. "Population and Resources of Cape Cod," 78.
126. Ibid.
127. Sheryl A. Lajoie and Brian A. Lajoie, "Sandwich Downs," East Sandwich Massachusetts, November 15, 2007. Located in Cape Cod Community College Library, Nickerson Room.

128. Nickerson, *Days to Remember,* 9.
129. Faught, *Falmouth,* 130.
130. Ibid., 57. In 1940 a small cottage in Falmouth rented for $400–$600 a season while larger homes went for $1,000–$2,500 a season and a season rental of an estate would go for almost $5,000. In 1940 there were over 500 houses for rent with 150 of them renting for over $1,000 a season.
131. Kathy Peiss argues that this embrace of vigorous commercialized heterosocial activity was pioneered by young working-class women in the early years of the twentieth century and was then adopted by the middle class. See Kathy Lee Peiss, *Cheap Amusements: Working Women and Leisure in Turn of the Century New York* (Philadelphia: Temple University Press, 1986).
132. Even golf became more available to the middle classes as resorts that catered to this clientele began building small putting greens. By the 1920s these putting greens were being replaced by a series of very short putting lanes on artificial turf. These miniature golf courses went out of favor in the Depression only to be brought back as miniature golf in the 1940s and 1950s with obstacles added (windmills were particularly popular on the Cape).
133. Faught's survey of activities by summer visitors to Falmouth found that bathing and swimming were the most popular activities followed by boating and sailing. (Remember that Falmouth had a significantly higher population of the very wealthy than other Cape towns). Tennis, golf, and fishing closely followed boating and sailing in popularity. Faught, *Falmouth,* 47.
134. Chatham and Chatham, *Cape Coddities,* 29–35.
135. Ibid., 23.
136. Christopher Winsor, *Camping in Wellfleet* (self-published, 2009), 1, 2.
137. Ibid. Winsor's family sent their camping gear on ahead by Railway Express. Once in Wellfleet they would stop first at the rail depot to pick up their goods then go next door to Nickerson's Lumber and buy wood for tables, posts, and uprights before heading out to camp at Cahoon Hollow. After 1939, when the Railway Express service ended, they loaded their gear on top of the car or on a trailer they pulled along behind the car.
138. O'Connell, *Becoming Cape Cod,* 51, 52. This book provides an excellent discussion of the process of selling Cape Cod.
139. Henry C. Kittredge, *Cape Cod: Its People and Their History* (Boston: Houghton Mifflin, 1968), 309.
140. Thompson, *History of Plymouth, Norfolk and Barnstable Counties,* vol. 2, 819.
141. Quoted ibid., vol. 2, 850–51.
142. Quoted ibid.
143. O'Connell, *Becoming Cape Cod,* 52.
144. Faught, *Falmouth,* 135. The Advancement Plan lasted for only two years before going out of existence.
145. Ibid., 98.
146. Lincoln, *Cape Cod Yesterdays,* 23.

147. Berger, *Cape Cod Pilot,* 124.
148. In 1935 the WPA estimated that more than 175,000 visitors came to the Cape for vacation and in 1936 state troopers counted 55,000 cars crossing just one of the Cape bridges on a single Sunday. Berger, *Cape Cod Pilot,* 3.
149. Ironically, despite the diking and drainage of the Herring River estuary, the cost of mosquito control only increased in Wellfleet. Beginning in 1912 the town of Wellfleet spent significantly more money on mosquito control than before the dike. Town of Wellfleet, *Annual Reports, 1912–1934,* Wellfleet Town Library. The Cape Cod Mosquito Control Project spearheaded much of this activity on the Cape between 1910 and 1960. Nickerson, *Days to Remember,* 277.
150. "Population and Resources of Cape Cod," 79.
151. John Portnoy and Michael Soukup, "From Salt Marsh to Forest: The Outer Cape's Wetlands," *The Cape Naturalist* 11 (Fall 1982), 29–34.
152. Faught, *Falmouth,* 82.
153. Ironically, the Cape's many small harbors encouraged the emergence in the 1920s of a new Cape industry: bootlegging. Moving alcohol from off-coast ships coming down from Nova Scotia and other maritime provinces to the Cape, and then on to Boston or New York, provided a significant boost to local employment. The Cape also sported several well-known speakeasies, one of the most impressive being the Casa Madrid in Yarmouth. The repeal of the Volstead Act in 1933 ended this lucrative job opportunity.
154. The Coonamessett Ranch, occupying thousands of acres, was an attempt to bring cattle raising back to the Cape. The ranch also hoped to become a model for successful large-scale agriculture on the Cape. After struggling several years the ranch finally collapsed with the coming of the Depression.
155. Faught, *Falmouth,* 34; O'Connell, *Becoming Cape Cod,* 62.
156. Teller, *Cape Cod and the Offshore Islands,* 31. By 1949 there were 1,500 miles of macadam roads on the Cape.
157. South Wellfleet became home to the Austin Ornithological Research Station (now a Massachusetts Audubon Society center) in the late 1920s when Dr. Oliver Austin bought up an abandoned asparagus farm.
158. The sandy peninsula around Skagen in the north of Denmark saw a similar type of artists community develop at roughly the same time.

7. THE GOLDEN AGE OF TOURISM

1. James C. O'Connell, *Becoming Cape Cod: Creating a Seaside Resort* (Hanover, NH: University Press of New England, 2003), 99.
2. See Adam Rome, *Bulldozer in the Countryside: Suburban Sprawl and the Rise of American Environmentalism* (Cambridge, UK: Cambridge University Press, 2010) for a discussion of postwar suburbanization.
3. In 1950 Boston's population at 801,000 was the highest in its history, but by 1980 it had fallen to 563,000, while the metropolitan region grew from 2.2 million in 1950 to 2.9 million by 1980.

4. The area housed the nation's leading universities and research centers, particularly Harvard University and MIT. Cultural institutions that added to the quality of life and medical centers created a context that encouraged innovation and a community that became an incubator for new enterprises.
5. In recent years there has been an interest in what has been called exurbia. These are settled areas constructed beyond the traditional commuting range whose homeowners either work at home via computer or commute into work on an irregular basis. The Cape has experienced this phenomenon, especially on the Upper Cape, but across most of the Cape it is retirees and vacation homeowners who drive the housing market.
6. In the early twentieth century the wealthy sought out the Cape for its beauty, seclusion, and the natural amenities that would enhance their life style. With the coming of the automobile, middle- and upper-middle-class visitors came looking for quaint settings, odd characters, and old-fashioned Yankee communities. In the postwar years more and more visitors to the Cape were looking for beaches on which to play or to appreciate the esthetics of its natural beauty. This shift is reflected in the popular literature of the Cape. In the first half of the century popular Cape books focused on its quirky old Yankee inhabitants and communities, while postwar literature of the Cape reflected a fascination with nature. For an example of the older Cape literature see Henry C. Kittredge, *Cape Cod: Its People and Their History* (Boston: Houghton Mifflin, 1930); Katharine Dos Passos and Edith Shay, *Down Cape Cod* (1974); Joseph C. Lincoln, *Cape Cod Yesterdays* (New York: Blue Ribbon Books, 1935); Lincoln, *All Along Shore* (New York: Coward McCann, 1931); Mary Rogers Bangs, *Old Cape Cod: The Land, the Men, the Sea* (Boston: Houghton Mifflin, 1920); Dennis and Marion Chatham, *Cape Coddities* (Boston: Houghton Mifflin, 1920); Samuel Chamberlain, *Cape Cod In the Sun* (New York: Hastings House, 1937); Katherine Crosby, *Blue-Water Men and Other Cape Codders* (New York: Macmillan, 1946); and Agnes Edwards, *Cape Cod New and Old* (Boston: Houghton Mifflin, 1918). For the increased focus on the Cape as natural place see: John Hay, *The Great Beach* (Garden City, New York: Doubleday, 1963); Hay, *The Undiscovered Country* (New York: W. W. Norton, 1981); Hay, *Nature's Year: Seasons of Cape Cod* (Garden City, New York: Doubleday, 1961); Hay, *The Run* (Garden City, New York: Doubleday, 1959); Clare Leighton, *Where the Land and Sea Meets the Sea: The Tide Line of Cape Cod* (New York: Rinehart and Co., 1954); and more recently Robert Finch's books *Special Places on Cape Cod and the Islands* (Beverly, MA: Commonwealth Editions, 2003), *The Primal Place* (New York: W. W. Norton, 1983), and *Common Ground: A Naturalist's Guild to Cape Cod* (Boston: David R. Godine, 1981).
7. From the 1950s through to the end of the century, deaths outpaced births. The Cape's dramatic population growth, more than 50 percent in the 1950s and 1970s and more than 26 percent in the 1980s and 1990s, was driven by in-migrants. The overwhelming majority of immigrants to the Cape came from other areas of New England. Marilyn Fifield and Cape Cod Commission, "Cape Trends: Demographic and Economic Characteristics and Trends, Barnstable County, Cape Cod," 5th ed. (Barnstable, MA: Cape Cod Commission, 1998).
8. O'Connell, *Becoming Cape Cod*, 127. In 1985 only 15 people in Eastham were employed in fishing and agriculture whereas 150 were employed in construction. Fifield, "Cape Trends."

9. O'Connell, *Becoming Cape Cod*, 127.
10. The majority of the Cape's working population found employment in services and government. Fifield, "Cape Trends."
11. Army Corps of Engineers, New England Division, Massachusetts Coastal Zone Management, "Cape Cod Easternly Shore Beach Erosion Study," Appendix 1, F-2, Appendix 3, F-3, 4.
12. Ibid.
13. In 1990 22 percent of Cape residents were over 65 compared to only 14 percent in the state. "Cape Trends."
14. O'Connell, *Becoming Cape Cod*. These summer home owners contributed 11 percent to the Cape's economy.
15. Fifield, "Cape Trends."
16. As late as 1950 Wellfleet had more than five grocery stores offering food and dry goods. By the 1970s the town was down to two plus an A&P open only during the summer. The expansion of the Stop and Shop in Orleans subsequently led to the closing of the A&P, leaving the town with two year-round grocery stores.
17. By 1980 fewer than 1 percent of Cape residents were employed in agriculture. Fifield, "Cape Trends."
18. James Nickerson, "The Association for the Preservation of Cape Cod and the Natural Resources to be Preserved" (Orleans, MA: APCC, January 1977), 13, 14.
19. These new homes were typically small two-bedroom homes with a porch sometimes attached to the front. They were built over a crawl-space cinderblock foundation with a low roof. Some had roofs with enough pitch to add a small low-ceiling room upstairs. Almost all were covered with shake shingles.
20. O'Connell, *Becoming Cape Cod*, 112.
21. The ban, ironically, led to a building boom, as any plans filed before the ban went into effect were allowed under a "grandfather" zoning clause. The town saw its motel rooms increase by 20 percent. O'Connell, *Becoming Cape Cod*, 101.
22. Army Corps of Engineers, "Cape Cod Easternly Shore, Beach Erosion Study," vol. 1, 9, 10.
23. Ibid., Appendix 3–4.
24. The 2000 census found 22.9 percent of the population to be over 65.
25. Dorothy Sterling, "Our Cape Cod Salt Marshes," APCC Information Bulletin #6 (Orleans, MA, 1976), 26.
26. Army Corps of Engineers, "Cape Cod Easternly Shore, Beach Erosion Study," 9. By the end of the century over 60 percent of the homes in Wellfleet belonged to nonresidents.
27. O'Connell, *Becoming Cape Cod*, 132.
28. Fifield, "Cape Trends."
29. Marcia Monbleau, *Pleasant Bay: Stories from a Cape Cod Place* (Harwich, MA: Friends of Pleasant Bay, 1999), 3.
30. Although the Mid-Cape Highway relieved some of the traffic congestion on Route 28, the completion of the Southeast Expressway down to the Sagamore Bridge in 1963,

which cut the travel time between Boston and the Cape, dramatically increased the flow of cars onto Cape highways.
31. Institute of Man and the Environment, "Massachusetts Coastal Zone Management Program and Final Impact Statement" (Boston, 1978), 3.
32. In its 2011 report, one of the major concerns of the Cape Cod Commission was the "challenge to find affordable housing." Cape Cod Commission, "Coastal Use Templates of Economic Development," May 2011, 1.
33. See O'Connell, *Becoming Cape Cod*, 112, for a discussion of the debate over zoning and planning on the Cape.
34. By 2000 the Cape had a density of 562 persons per square mile, while the corresponding state figure was 810. In contrast, the Cape in 1930 had 82 persons per square mile while the state had 272. From 1950 to 2000 the Cape grew by more than 400 percent.
35. Quoted in O'Connell, *Becoming Cape Cod*, 113.
36. Herbert Whitlock and James Nickerson were leading founders of the Association for the Preservation of Cape Cod.
37. Nickerson, "The Association for the Preservation of Cape Cod," 2.
38. Sterling, "Our Cape Cod Salt Marshes," 13; Cape Cod Commission, "Coastal Use Template."
39. Although James C. O'Connell does not fall victim to this reductionist fallacy, he does provide evidence of its widespread acceptance. See O'Connell, *Becoming Cape Cod*, 98, 99.
40. Sterling, "Our Cape Cod Salt Marshes," 26; APCC, "The Action Required Now to Protect the Quality of Cape Cod's Drinking Water" (February 1979), 15.
41. To this day, land on the Upper Cape purchased by the towns to protect their water supply far exceeds private trust land and land purchased since the 1999 establishment of the Land Bank. See, for example, The Dennis Conservation Trust, "Dennis Lands Conserved by DCT and Others," dennisconservationtrust.org.
42. O'Connell, *Becoming Cape Cod*, 113.
43. "Celebrating 50 Years of Cape Cod National Seashore," *Provincetown Banner*, May 19, 2011.
44. O'Connell, *Becoming Cape Cod*, 113.
45. Henry Beston, *The Outermost House: A Year of Life on the Great Beach of Cape Cod* (New York: Henry Holt, 1928). Those familiar with the many town and National Seashore backside beaches available to visitors today would be surprised in reading Beston how barren and isolated the area was ninety years ago. Beston's house was incorporated into the National Seashore, but washed into the sea in the Northeaster storm of 1978.
46. With the exception of Dennis, which felt that since it had spent a great deal of its public money on land acquisition, the Outer Cape towns should do the same, most Cape towns other than the three most directly affected (Eastham, Wellfleet, and Truro) supported the park.
47. Charles H. W. Foster, *The Cape Cod National Seashore: A Landmark Alliance* (Hanover, NH: University Press of New England, 1985) 4, 5, 10, 11.

48. Although the park was supposed to limit development on park land, several persons used the interval between the passage of the park bill and the formal taking of land for the park to build homes in what became park land. Because of the ambiguity of the situation, the park offered these homeowners life tenancy in their property.
49. Cape Cod Commission, "History," www.capecodcommission.org.
50. Ibid.
51. Ibid. The participants in "Prospect: Cape" came to focus on diverse areas of concern. They listed environmental quality, land use, economic development, support services, health and human services, and housing. The initiative produced 60 recommendations and established a goal of putting public policy proposals into action over the next five years.
52. Cape Cod Commission, "History."
53. Institute of Man and the Environment, "Massachusetts Coastal Zone Management Program and Final Impact Statement," 3.
54. Army Corps of Engineers, "Cape Cod Easternly Shore V Beach Erosion Study," Appendix 1, F-40.
55. O'Connell, *Becoming Cape Cod*, 108.
56. Ibid.
57. Cape Cod Commission, "History."
58. Ibid.
59. O'Connell, *Becoming Cape Cod*, 108. The Act granted the Commission the "authority to prepare and oversee the implementation of a regional land use policy plan for the Cape, to recommend for designation specific areas of Cape Cod as districts of critical planning concern requiring special protections and to review and regulate developments of regional impact such s developments that meet minimum criteria pertaining to size, location or character and affect more than one municipality." See Cape Cod Commission, "History."
60. Dona Brown, *Inventing New England* (Smithsonian: Washington, DC, 1995), 216.
61. Cape Cod Commission, "Regional Policy Plan," www.capecodcommission.org.
62. Lawyers specializing in fighting against town building restrictions found ready customers across the Cape. The local Conservation Commissions, for example, which were empowered to protect wetlands, were made up of local non-expert citizens. Lawyers for the developers and builders would inundate these committees with facts. The developers would come in with their most extreme position; if the committee showed signs of rejecting the project, the lawyers would request a postponement. This process could go on for months with the lawyers for the developers bringing back slightly modified plans that would address some of the concerns of the committee, but never give the committee the opportunity to actually vote on the proposal. Eventually the lawyers hoped to wear down the committee and get an okay for a more modified version.
63. This act was modified in 2006 with the Community Preservation Act to allow funds to also be used for non–open land preservation expenditures.

64. The town of Barnstable, for example, had by the end of 2008 (the Land Trust's 25-year anniversary) more than 745 protected acres.
65. In 1951, 42,257 acres of the Cape was in agriculture or was open land, 45,065 were wetlands, and over 19,000 were either urban residential or for commercial use. Army Corps of Engineers, "Cape Cod Easternly Shore, Beach Erosion Study," vol. 2, 1979, F-7.
66. The reforesting within the National Seashore has led to a campaign of selective burning to try and maintain open land.
67. The late twentieth-century forest consisted mostly of pitch pine, scrub oak, and black locust.
68. Robert W. Eberhardt, David R. Foster, Glenn Motzkin, and Brian Hall, "Conservation of Changing Landscapes: Vegetation and LND-Use History of Cape Cod National Seashore," *Ecological Applications* 13 (2003): 68–84.
69. By 1975 85 percent of the Cape's economy was based on summer payrolls as a result of summer visitors. After trade and services came construction as the third heaviest source of employment. Army Corps of Engineers, "Cape Cod Easternly Shore V Beach Erosion Study," vol. 2, 1979, F-7.
70. Ibid., vol. 2, Appendix 1, F-8.

8. PROBLEMS IN PARADISE

1. John Portnoy and Michael Soukup, "From Salt Marsh to Forest: The Outer Cape's Wetlands," *The Cape Naturalist* 11 (Fall 1982): 29–34. Justification for the wetland drainage was linked to public health and mosquito control. Ditches were dug three feet below the surrounding marsh surface, and water drained out to sea. By the 1950s Barnstable County had over 1,000 miles of drainage ditches.
2. Carl Carlozzi et al., "Ecosystems and Resources of the Massachusetts Coast," Massachusetts Coastal Zone Management, Institute for Man and Environment, 15.
3. Dorothy Sterling, "Our Cape Cod Salt Marshes," APCC Information Bulletin #6 (Orleans, MA: 1976), 12.
4. Donald A. Crane for the Massachusetts Water Resource Commission, "Coastal Flooding in Barnstable County–Cape Cod, Massachusetts" (Boston: December 1962), ii.
5. The largest gap between hurricanes hitting the Cape occurred between 1972 and 1991, during which the Cape experienced a boom in coastal construction.
6. Cape Cod's shorefront had not become a major attraction before the second half of the twentieth century, so the damage from the 1938 storm on the Cape was limited in comparison to the immense damage and considerable loss of life in Rhode Island, where vacation homes had already taken up much of the shorefront, but only limited damage on the Cape. The hurricanes of August 31 and September 11, 1954 caused much more property destruction and damage.
7. See Theodore Steinberg, *Acts of God: the Unnatural History of Natural Disaster in America* (New York: Oxford University Press, 2000) for a discussion of the habit of building in areas prone to problems. This situation reoccurred recently after Hurricane Sandy when New Jersey used its influence to push rebuilding money through Congress with-

NOTES TO PAGES 176–182

out addressing the problem of building on barrier beaches or seeking funds to buy out beachfront property owners.
8. Crane, "Coastal Flooding," 33.
9. Ibid.
10. Ibid., 33, 34.
11. Ibid., 34, 35.
12. Ibid., 36.
13. Ibid., 44. In Massachusetts, when the private insurance companies refused to insure property close to the water, the state stepped in with a state-supported insurance program.
14. Ibid., ii.
15. Continued coastal development though the 1960s and 1970s led to massive destruction in the Blizzard of 1978, which destroyed when homes and structures across the Massachusetts coast. In response to that storm the state created the Massachusetts Coastal Zone Management to try to protect natural barrier beaches and wetlands. In 1980 the governor issued Executive Order #181 prohibiting the use of state funds to encourage growth and development on barrier beaches.
16. Sterling, "Our Cape Cod Salt Marshes," 26.
17. Wetlands had been considered wasteland and/or health hazards for much of the eighteenth, nineteenth, and early twentieth centuries. The new science of ecology encouraged scientists to relook at the role of wetlands in the health of the larger ecosystem. By the 1970s a significant body of scientific literature had been developed showing the importance of wetlands for both flora and fauna. This information helped bring fishermen, hunters, and birders into an uneasy alliance to protect wetlands.
18. Sterling, "Our Cape Cod Salt Marshes," 30–32.
19. APCC [Association for the Preservation of Cape Cod], "Protecting Our Lakes and Ponds: Legal and Environmental Considerations," February 1973.
20. Harvard University School of Public Health, "Survey of the Health Needs of Barnstable County" (Author: Cambridge, 1957), also known as the "Harvard Study."
21. CCPEDC [Cape Cod Planning and Economic Development Commission] and Environmental Protection Agency, "Water Quality Management Plan: Environmental Impact Statement for Cape Cod," vol. 2, 1978, 24, 25.
22. James McCann, "An Inventory of the Ponds, Lakes, and Reservoirs of Massachusetts–Barnstable County," Water Resources Research Center, University of Massachusetts, Amherst, 1969. 6.
23. APCC, "Protecting Our Lakes and Ponds."
24. APCC, "The Action Required Now to Protect the Quality of Cape Cod's Drinking Water," February 1979.
25. CCPEDC, "Water Quality Management Plan," vol. 1, 1978, 4.
26. James C. O'Connell, *Becoming Cape Cod: Creating a Seaside Resort* (Hanover, NH: University Press of New England, 2003), 134.
27. Environmental Management Institute, "Water Quality Assessment: Cape Cod" prepared for the Area Wide Waste Water Management Program, Cape Cod, Planning and Economic Development Commission, Barnstable County, 1976.

28. CCPEDC, "Water Quality Management Plan," vol. 1, 4.
29. In 2011 the town launched an initiative to rebuild the oyster reefs in part of the harbor in hopes that the oysters would filter nitrates out of the harbor water and protect the other beds.
30. The EPA designation meant that local towns could no longer ignore the issue of regional cooperation to protect ground water.
31. CCPEDC, "Water Quality Management Plan," vol. 2, 75. Cape Cod's average was significantly higher than the overall average for eastern Massachusetts. The large quantity of water used by restaurants, motels, laundries, and car washes accounts for this extra water use.
32. State law prohibits the discharge of treated sewage into the ocean off the Cape. Most Cape Cod waste treatment facilities by the mid-1980s did only secondary treatment. See "Cape Cod's Lifeblood is Endangered," *Boston Globe,* November 25, 1985.
33. James Nickerson, "The Association for the Preservation of Cape Cod and the Natural Resources to be Preserved" (Orleans, MA: APCC, January 1977).
34. Pollution such as toxic wastes entering the aquifer pollutes the water immediately surrounding its entry point. It slowly migrates out from that point following the natural flow of the ground water. Although this keeps the pollution localized it also takes far longer for the pollution to flush out of the system. Pollution dumped fifty years ago at Otis Air Base is still slowly moving through the aquifer.
35. The town dump in Eastham leached toxic materials into the wells of the surrounding homes. A vote before the town meeting to create a municipal water system for those homes failed. The town opted instead to supply bottled water to those affected.
36. Environmental Management Institute, "Water Quality Assessment: Cape Cod."
37. CCPEDC, "Water Quality Management Plan," vol. 1, 4.
38. Data from capecodcommission.org. Public water sources account for 9.3 billion gallons of water used, while private wells account for 1.2 billion gallons.
39. Saltwater infusion in Provincetown's water led it to establish wells in Truro. When a gas station spill in 1977 shut down one of Provincetown's two remaining well fields, the town again turned to Truro, which this time resisted allowing Provincetown to drill more wells in the town. Provincetown then got an emergency well from the CCNS. In 1977 its North Truro well registered significant saltwater infusion. CCPEDC, "Water Quality Management Plan," vol. 2, 75, 77.
40. The development of a Cape regional waste disposal system is under discussion, but these systems also raise problems for the Cape's aquifer. With a withdrawal rate of 9.3 billion gallons of water from the aquifer by public water systems, if that much water is treated in a treatment facility and then released into the bay or ocean, that will lower the available fresh water in the aquifer.
41. CCPEDC, "Water Quality Management Plan," vol. 1.
42. By 2010 individual (traditional Title V) systems on the Cape were processing over 60 million gallons of waste per day. These individual systems did a poor job of eliminating nitrate wastes.
43. CCPEDC, "Water Quality Management Plan," vol. 2, 75.

NOTES TO PAGES 185–191

44. Ibid., vol. 3, 15.
45. Joel Tarr, *The Search for the Ultimate Sink: Urban Pollution in Historical Perspective* (Akron, OH: University of Akron Press, 1996).
46. APCC, "The Action Required Now," 15. Wellfleet, for example, closed its dump in 1990 and began transferring its wastes to off-Cape sites.
47. In the mid-1980s all but one of the towns on the Cape signed on to a solid waste disposal facility near Wareham, Massachusetts, which would take the Cape's solid waste and burn it.
48. Nickerson, "The Association for the Preservation of Cape Cod and the Natural Resources to be Preserved."
49. See Elizabeth Pillsbury, "'All at Last Returns to the Sea:' Land Use and Water Quality on Southern New England's Shore," in *A Landscape History of New England*, ed. Blake Harrison and Richard W. Judd (Cambridge: MIT Press, 2011), 192.
50. The Save the Dunes Council pushed for a national park on Lake Michigan's southeastern shore. The council then joined with other Lake Michigan organizations to push for legislation to protect lakeshores. Their activity linked with others in pushing Congress to address the issue of the degradation of coastal zones. See John T. Cumbler, *Northeast and Midwest United States: An Environmental History* (Santa Barbara: ABC-CLIO, 2005), 188–90.
51. The EPA was charged, among other things, "to conduct investigations, studies, research and analysis relating to ecological systems and environmental quality." Congress passed the Coastal Zone Management Act (CZMA) in response to growing public pressure over the challenge of continued growth in the coastal zone. The act, administrated by the National Oceanic and Atmospheric Administration's (NOAA) office of Ocean and Coastal Resource Management (OCRM), provided for the management of coastal resources in order to balance economic development with environmental conservation. CZMA defined its overall program as aiming to "preserve, protect, develop, and where possible, to restore or enhance the resources of the nation's coastal zone" (coastalmanagement.noaa.gov).
52. Institute of Man and the Environment, "Massachusetts Coastal Zone Management Program and Final Impact Statement" (Boston: Author, 1978), 3.
53. Ibid.
54. Wellfleet passed its first building regulations in 1957, establishing lot sizes of 10,000 square feet. In the 1970s Wellfleet increased its lot size to 20,000 square feet.
55. APCC, "Protecting Our Lakes and Ponds."
56. Ibid.
57. DDT (dichlorodiphenyltrichloroethane) was first used domestically against the Japanese beetle and then against ticks. Along with malathion and parathion, DDT became the central weapon against insects in gardens, lawns, and homes. In 1962 Rachel Carson serialized her now famous *Silent Spring* in the *New Yorker* and launched the first major attack against the indiscriminate use of pesticides.
58. Quoted in Stephan Mulloney, *Traces of Thoreau* (Boston: Northeastern University Press, 1998), 24.

59. "Water Quality Management Plan: Environmental Impact Statement for Cape Cod," vol. 2, 93.
60. David S. Blackmar, "A Frightening Array of Household Chemicals," *The Cape Naturalist* 12 (Winter 1983): 44–45.
61. The state passed water quality legislation to protect waterways, water aquifers, and fresh and salt water from harmful discharge, MGL c.21A, s.13. Section 11C reads: "The commissioner . . . may . . . for the purpose of promoting public safety, health and welfare and protecting public and private property wildlife, fresh water fisheries and irreplaceable wild, scenic and recreational river resources, adopt . . . orders regulating, restricting or prohibit dredging, filling, removing or otherwise altering or polluting the scenic and recreational rivers and streams of the commonwealth. Section 13 reads: "The commissioner of the department of environmental protection shall adopt and from time to time amend, regulations to be known as the state environmental code." "Local boards of health shall enforce said code" but if the local boards fail to act, "the department may in like manner enforce said code." The Commonwealth of Massachusetts, "Massachusetts General Laws, c.21A, s.13." In 1995 this law was challenged in a court case from Bourne, *Tortorella v. Board of Health of Bourne*, 39 Mass. Appeals Ct, 277. The court upheld Title 5.
62. CCPEDC, "Water Quality Management Plan," vol. 3, 15.
63. Until 1980 there were open salt piles in both Harwich and Truro.
64. Ben McKelway, "Roads in the Snow—Hold the Salt," *The Cape Naturalist* 11 (Spring 1983): 69.
65. Ibid.
66. From 1955 to 1969 U.S. Air Force planes dumped between 1 and 5 million gallons of aviation fuel on the base. In addition to these, solvents, degreasers, photographic chemicals, and other toxic chemicals were simply poured into dry wells.
67. CCPEDC, "Water Quality Management Plan," vol. 2, 75.
68. Love Canal was only one of several toxic hazardous waste issues that galvanized public action. In 1979 the EPA created the Hazardous Waste Enforcement Task Force to locate and begin action on toxic sites. By 1980 the HWETF had located 9600 potential sites. In the fall of 1980 Congress passed the Comprehensive Environmental Response, Compensation and Liability Act (Super Fund Act). By the early 1980s the base at Otis was added as a Super Fund site.
69. In the 1980s the Massachusetts DEQE closed Falmouth's Ashumet well and provided bottled water to local residents when testing revealed contamination from the sewage treatment plant on the Massachusetts Military Reservation had infiltrated the aquifer from which water was drawn.
70. Geological studies indicate that the natural activity of wave and wind is slowly moving the Cape, eliminating the Outer Cape. Global warming will add to this process as water levels rise. These rising water levels will flood low-lying areas and push up water levels into the leach fields of existing septic systems.
71. The state built a dike to stop the flow of sand from East Harbor into Provincetown Harbor, creating a beach. Water flowing around the tip of the Cape pulls sand from

south of the old opening of East Harbor and deposits it to the north increasing the beach there and making the water shallower for boating. Beachfront residents south of the opening found they lost 6 to 10 feet of beachfront a year.
72. Quoted in Graham Giese and Rachel Giese, *Eroding Shores of Outer Cape Cod* (Orleans, MA: AAPC, October 1974).
73. Ibid.

CONCLUSION

1. Gustavus Swift, the famous Chicago meat packer, was a Cape Codder who early in his career sent Cape beef to Boston.
2. See Carroll Wright, Bureau of Statistics of Labor, *Census of Massachusetts 1885*, vol. 3, "Agricultural Products and Property" (Boston: Wright and Potter Printing Company, 1887), 7–26 for evidence of declining emphasis on beef, wool, and cereals and the growing importance of vegetables, fruits, dairy products, and poultry.
3. Commonwealth of Massachusetts, Department of Labor and Industries, "Population and Resources of Cape Cod" (Boston: Wright and Potter Printing Company, 1922), 12.
4. The town of Wellfleet still had 87 milk cows as late as 1900. Even in 1920 it had 74 cows. With the coming of the automobile the number of horses in Wellfleet declined dramatically. In 1900 there were 112 horses, almost one for every family; by 1920 there were exactly half as many, 56. Town of Wellfleet, *Annual Reports of the Officers of the Town of Wellfleet*, 1900, 1920.
5. Anthony L. Marshall, *Truro, Cape Cod, As I Knew It* (New York: Viking Press, 1974).
6. For a discussion of the mackerel fisheries see Matthew McKenzie, *Clearing the Coastline: The Nineteenth Century Ecological and Cultural Transformation of Cape Cod* (Hanover, NH: University Press of New England 2010). McKenzie provides an excellent discussion of the conflict between the near-shore and offshore mackerel fishermen and their attempts to establish blame for the declining numbers of mackerel. In all probability both types of fishermen were stressing the mackerel stock.
7. See Richard Conniff, "In Search of the Real Cape Cod," *Yankee,* August 1998, 69; Mark Kurlansky, *Cod: A Biography of the Fish That Changed the World* (New York: Walker and Co., 1997), 221–23. With lower capital intensity and thus lower debt, Chatham cod fishermen held on, continuing to long-line for cod off of smaller boats.
8. William Birchen and Monica Dickens, *Cape Cod* (New York: Viking Press, 1972), 61.
9. Henry David Thoreau, *Cape Cod* (New York: Penguin Press, 1987; orig. 1865), 75. Despite his description of the Cape as barren, Thoreau admitted "I never saw an autumnal landscape so beautifully painted as this was" (225). See Hal K. Rothman, *Devil's Bargains; Tourism in the Twentieth-Century American West* (Lawrence: University of Kansas Press, 2000) for a discussion of the costs of tourism.
10. Marshall, *Truro,* 101.
11. As early as 1920 commentators were noting that the Cape was being transformed with "older homes purchased and renovated to fit the needs of city people." Dennis and Marion Chatham, *Cape Coddities* (Boston: Houghton Mifflin, 1920), 136–39.

12. The Cape's population doubled from the 1960s to 2000, by which time there were over 185,000 year-round residents and over 500,000 on a good summer weekend. Typical of this process was Wellfleet. In 1920 Wellfleet had 494 homes and the value of an acre of real estate was $105.78. In an economy where the average skilled worker earned $1,250.00 a year and most mortgages were short-term (four to eight years), this represented a significant but possible expenditure. Tourism and vacation homes came to Wellfleet in the 1920s. By 1930 the town had over 642 dwellings, a 75 percent increase. It still had 41 cows but now only 13 horses. The increased interest in Wellfleet as a place to vacation pushed up the value of real estate to $218.14 an acre, more than double what it was at the beginning of the decade, a decade in which real wages for working Americans did not grow significantly. The Depression reduced the upward pressure on Wellfleet's real estate values: property in 1940 was valued at $275.00 an acre. The town still supported 23 cows but had only 2 horses. The rugged beauty of Wellfleet's coast and its picturesque harbor with decaying old fishing boats drew the interest of those who wanted to own a piece of natural beauty. Two hundred and forty-seven new homes were built during the 1940s. Two cows remained in town, but the number of horses rose to 5, mostly for recreational riding rather than for work. The value of property rose to $321.45 an acre in 1950, higher than in the 1920s and 1930s but still within the reach of those who had a place within the productive economy. The trend beginning in the late 1940s continued into the 1950s. By 1960 there were over 3,040 dwellings in the town and real estate value was at $568.73 an acre. In 1970 the value of real estate had skyrocketed to $6,063.64 an acre and the town now had 50 riding horses. In 1980 real estate values had risen to $33,935.29 an acre. Town of Wellfleet, *Annual Reports of the Officers of the Town of Wellfleet*, 1920, 1930, 1940, 1950, 1960, 1970, and 1980.
13. The state Wetlands Protection Act (which prohibits any filling, excavation, dredging, or other alterations of salt- and freshwater wetland areas, and as well protects the banks, dunes, beaches, vernal pools and land under designated waters) empowered the town Conservation Commissions to oversee any building project that could have a wetland impact. In their initial years these commissions did little to restrain development, but over the last twenty-five years they have become more aggressive in protecting wetlands.
14. Contrary to early projections, the Cape's population growth has dramatically slowed, even gone into reverse in some places. Between 2000 and 2010 the Cape's population stood at 215,888, well below the 270,000 predicted. By the year 2010 the year-round population of the Cape declined by 2.97 percent. The loss of population came mostly from families with children, with the 0–17 year olds decreasing in number by 19 percent. The high price of housing is one of the major issues driving away young families. "Barnstable County," quickfacts.census.gov.
15. Cape Cod Commission, "Resource Center," www.capecodcommission.org.
16. Wellfleet, Truro, and Orleans have all had major battles over the practice of buying small cottages, tearing them down, and building larger and far more expensive structures. Although town conservation commissions and citizen groups have complained

NOTES TO PAGES 203–204   261

about this practice, they have had limited success in battling it. See the conflict over the "abominable house" in Wellfleet. The Cape Cod *Banner* carried many stories on these conflicts between 2006 and 2010.
17. Orleans Planning Department.
18. Outer Cape towns like Wellfleet are facing a severe crisis in affordable housing. The average sale price of a home in Wellfleet in 2011 was $525,900. Town of Wellfleet, *Annual Town Report,* 2011, 80.
19. Ibid.
20. In an attempt to control this process the towns of Truro and Wellfleet have introduced zoning changes that would require special permitting processes for tearing down and building a significantly larger structure, or overloading a lot with a large structure. Wellfleet's measure, although supported by the selectmen, the local comprehensive plan implementation committee, and the planning board, was voted down by the town meeting. Town of Wellfleet, *Annual Town Report,* 2011, 44, 80, 179.
21. Towns with the largest number of homes packed tight together have already abandoned individual septic systems for town treatment plants. Provincetown, for example, has a town system whereas Wellfleet has some 3,000 on-site septic systems. The upfront cost of developing a town-wide system is high, leading to significant resistance; but once established, a central system costs only $500 to remove a pound of nitrogen whereas the cost for an on-site system is $1,000 per pound of nitrogen.
22. James O'Connell, *Becoming Cape Cod: Creating a Seaside Resort* (Hanover, NH: University Press of New England, 2003), 135, 99.
23. Small, moderately priced motels added to the Cape's supply of cottage colonies in the late 1950s. These motels made rooms available for those without the resources for renting a full cottage for a week or more. James Pendergast built the Cape's first motel, a simple four-room motor court, in 1939. In 1949 he built three more in Hyannis and Craigville. By the 1950s more motels were thrown up, especially along Route 28. The combination of ease of automobile access to the Cape and moderately priced accommodations encouraged a more diverse visiting and settling population on the Cape than the far more exclusive and expensive vacations spots such as Martha's Vineyard, Nantucket, or Newport. See O'Connell, *Becoming Cape Cod,* 99.
24. Even a 50-year-old trailer in a trailer court could cost $220,000 by the beginning of the twenty-first century. *The Week* magazine, in an article on "Homes on Cape Cod" (all but one in the seven figures), listed one small simple three-bedroom ranch inland from the coast and two blocks from a pond as a "steal of the week" at $299,000. Not only are housing prices rising dramatically but so are rental costs. Many of the small cottages that sprang up in the first half of the century were either torn down or upgraded and sold as condominiums. A single week on the Cape in a modest cabin or home can now cost as much as $1,500 or more. Although the Cape still offers inexpensive camping in Nickerson State Park, in the Audubon Sanctuary, and in the few scattered campgrounds still available on the Cape, these campsites are in ever-greater demand.
25. The decline in construction on the Cape in the first decade of the twenty-first century was partly due to the general collapse of the economy, but also, on the Cape, to the limits

of open land for new construction. Those who found employment in the building trades now find themselves in a double bind: less work and ever more expensive housing costs.
26. Conniff, "In Search of the Real Cape Cod," 54.
27. The Cape after experiencing dramatic growth over the last 3 decades of the twentieth century experienced a drop in population from 2000 to 2010. This 2.4 percent drop in population compares to growths of 49 percent from 1950 to 1960, 38 percent from 1960 to 1970, and 49 percent from 1970 to 1980.
28. The Buzzards Bay Coalition and the Conservation Law Foundation sued the EPA for not enforcing regional groundwater protection despite its own finding that the Cape is served by a single aquifer. The suit is on hold while the county attempts to develop a regional sewage treatment system. In 2012 the Massachusetts Estuary Project established nitrogen limits for estuaries. It is expected that most towns will have to take even more aggressive steps, most likely more sophisticated central treatment systems, to meet the new standards.
29. These town committees usually work in combination with town housing authorities and town housing partnerships.
30. Using a number of revenue sources including Community Preservation Act money, Provincetown has just completed a major complex of affordable units and Wellfleet is in the midst of building an affordable housing development.
31. In a 1991 survey done by the Cape Cod Commission, a majority of Cape residents supported limited growth in the tourist industry and advocated the commission push light industry rather than tourism. Dona Brown, *Inventing New England: Tourism in the Nineteenth Century* (Washington, DC: Smithsonian Institution Press, 1995), 217.
32. Cape Cod National Seashore has over 5 million visitors a year, which ranks it as the country's fifth-busiest national park. Conniff, "In Search of the Real Cape Cod," 63.
33. The world of science has changed dramatically over the last three hundred years. The first few generations of white settlers on the Cape worked from a basic empirical world. When fish stocks seemed to be declining they acted to limit catches. When woods appeared overcut or grasses overgrazed they limited cutting and grazing. By the nineteenth century scientists were counting fish and tracking plant and animal species. The twentieth century has given us far more sophisticated scientific tools and approaches. The National Park Service now employs plant and animal biologists, ecologists, and geologists to understand natural processes. The interactions between plants and animals and climate are taken into consideration before actions are taken. Public health officials can now analyze water to determine oxygen content and trace the migrations of toxic and hazardous materials. With the aid of these new scientific tools and approaches we are in a much better position to make decisions that will protect the overall health of the environment. But better science and better knowledge do not always lead to better decisions. As we have seen in the debate over climate change, just because the science is available does not mean the public or decision makers will follow the scientifically rational choice. Fish biologists may be able to bring science to the table in a discussion about limiting fish take, but the tradition and politics that fishermen bring to the table often override the science.

34. Although the colony continued to hold the land in common, even within the town of Provincetown, it continued to struggle with problem of overgrazing the commons and cutting of woods from the dunes. The point here is not that the original settlers solved the problem of overuse, rather that they struggled to find a balance between the public good and private interest.
35. One of the actions proposed by the Cape Cod Commission is for the Cape to develop light high-tech industry. Yet Cape Codders still have to find a way to live with their environment and build a base of support for such a life. Cape Cod Commission, "Comprehensive Economic Development Strategy" (capecodcommission.org). Ironically one of the new high-tech options for the Cape, wind power, has created dramatic divisions within the conservation movement. Many Cape Cod conservationists and environmental organizations have endorsed the idea of wind power on the Cape as an environmentally friendly alternative to the fossil fuel power plant in Sandwich and a source of environmentally friendly jobs. Other conservationists oppose wind power as unsightly, potentially noisy, and disruptive of the Cape's traditional vistas. This conflict has been most dramatically fought out over the plans of Cape Wind to put wind turbines in Nantucket Sound.
36. Figures on the importance of aquaculture activity vary widely. This figure is for the state, but the majority of aquaculture takes place on Cape Cod. The state Division of Marine Fisheries reported that in 2012 the state's 345 aquaculturists landed 80,000 bushels of oysters, with a value of $8.6 million paid to the fishermen. Doug Frazer, "Local Oysters Blamed for Illness," *Cape Cod Times,* November 10, 2012.
37. Plans have also been put forward for developing oyster reefs to filter and clean the estuary waters of the Cape.
38. Due to the collapse of the fish populations off the coast of New England, the fishing industry of old will not return to the Cape. Provincetown's fishing fleet of 60 in the early 1980s was down to a dozen by the end of the century. Careful conservation efforts under pressure from federal fishing regulations should help in restoring fish stocks, but fishermen continue to rail against conservation actions. If the scientists and the fishing community can rally around a serious conservation effort, there is a chance the fish stock will rebuild, but it would have to be based on a much more limited catch, with the price of fish higher.
39. See Cape Cod Commission, "Regional Plans," www.capecodcommission.org.
40. See Brian Donahue, *Reclaiming the Commons: Farms and Forests in a New England Town* (New Haven: Yale University Press, 1999) for a discussion of the process of integrating active use of the land with a conservationist future.
41. At this point Robert Finch is probably the Cape's most renowned living nature writer. Ironically, although he lives on the Cape for most of the year, he summers in Newfoundland away from the Cape's summer crowds.
42. The passage of the Coastal Zone Management Act reflected a growing national awareness that the problems of pollution, overdevelopment, and access are national rather than purely local problems.
43. See Elizabeth Pillsbury, "'All at Last Returns to the Sea': Land Use and Water Quality

on Southern New England's Shore," in *A Landscape History of New England,* ed. Blake Harrison and Richard Judd (Cambridge: MIT Press, 2011), 181–98, for a discussion of the problems of development and pollution on Long Island Sound. As with the Cape, the population of the Long Island coast jumped dramatically over the last fifty years. With population growth came increased building of highways, parking lots, and shopping centers, which reduced the land's ability to filter pollution flowing into the region's waters. Builders, as on the Cape, looked to wetlands and marshes for cheap land for development. The erosion of beachfront property due to hurricanes such as Sandy brings forth media attention, but more damaging for the Outer Cape (and for the North Shore of Massachusetts and the New Hampshire and Maine coastal beaches) are winter Northeastern storms. These storms pound coastal areas with tremendous energy, washing away beaches and homes. It was a Northeaster of 1978 that washed away the "Outermost House" 53 years after it was built, as well as the beach house and parking lot of Coast Guard Beach in Eastham. The Northeaster of 1991, "the perfect storm," ripped across the Cape eroding beaches and sending homes into the surf. Although Hurricane Sandy destroyed homes and beaches from New Jersey to Connecticut, the Northeaster that followed a month later and the one that slammed ashore March 7, 2013, did more damage than Sandy on the Cape and North Shore. "It's Move It or Lose It in Path of a Nor'easter," *New York Times,* March 19, 2013.

44. Local authorities in Spain, Portugal, and Italy have been under pressure to overlook building restrictions and preservation issues in hopes of encouraging economic growth and employment. The consequences are readily apparent along the coast of Spain, where building flats have overrun the shorefront. In Italy, along the Sardinian coast, it has become common for developers to go ahead with a prohibited development project and afterward get a pardon. These coastal regions are also experiencing increased pollution and congestion as well as unsightly development.

INDEX

Page numbers in italics refer to illustrations

Agrey, Thomas, 43
alewives, 36, 219n7, 228–29n32, 229n38; fishing for, 22, 58, 59, 70; Native Americans and, 22, 59; and river obstructions, 48, 59, 70, 179
Altpeter, L. Stanford, 104
"American plan" for hotel costs, 104, 113, 125, 134, 142
American Revolution, 48, 51, 73, 81, 226n69
Ames, Fisher, 81
anadromous fish, 22, 36, 59, 219n8; spring runs of ("herring runs"), 4, 41, 84. *See also* alewives; river herring; shad
aquaculture, 60, 208, 229n43, 263n36
aquifer, 182–84, 185, 192–93, 256n40; single, for entire Cape, 182, 183; danger to, from saltwater infusion, 182–83, 184, 193–94; leaching of wastes into, 187, 194–95, 256n34, 258n69
Army Corps of Engineers, 172
artists, 126, 148, 249n158
Association for the Preservation of Cape Cod (APCC), 162, 181, 186–87
Atkins, Captain, 54
Austin Ornithological Station, 163
automobiles, 128–30, 152, 159–60, 168. *See also* highways; parking lots; traffic

Baker, Florence, 103

Baker, Isaiah, 86, 87
Baker, Lorenzo Dow, 1–4, 87, 105; and Wellfleet dike controversy, 1, 3–4, 22, 145, 178, 215n8; background of, 1–3, 72, 96–97, 187; and Chequessett Inn, 2–3, 97, 104, 124, 215n7; vision of, for Cape, 4, 5, 10, 96–98, 113, 187; and Methodist camp meetings, 97, 102, 113; cottages built by, 105, 125, 202, 203
Ballston Beach Colony (Truro), 105
banks. *See* fishing banks
Banneker, Garret A., 143–44
Barnstable, 45, 48, 71, 158, 165; natural setting of, 9–10, 17, 33, 36, 223n21; Native Americans in, 21, 36–37, 78; farming in, 21, 49, 52, 71, 82; in colonial era, 36–37, 222n14; forests and deforestation in, 46, 47, 171; maritime industries in, 54, 61, 63, 65, 66, 77, 83, 88, 89; transportation to and from, 61–62, 91, 159; water and sewage in, 180, 185, 190, 192. *See also* Centerville village; Cotuit village; Craigsville village; Hyannis; Marston Mills village; Osterville village
Barnstable County Agricultural Society, 82
barrel making, 60, 92–93, 198
bass, 49, 85, 99, 236n22

265

Bass River (Yarmouth), 103, 192; wharves on, 53, 61, 62, 85
Bates, Katherine Lee, 92
beaches, 46, 127, 190, 252n45; barrier, 17, 255n15; whales stranded on, 38, 41, *42*, 224n37; bathing, 98, 132, 156, *156;* fish processing on, 120, 121; on Outer Cape, 125, 164, 165, 166, 205, 209, 258n71; parking lots for, 159–60; crowding of, 159–60, 168; and Cape Cod National Seashore, 164, 165, 166, 205; on ponds, 181; erosion of, 195–96, 209, 252n45, 264n43. *See also* beach grasses
beachfront property, 101, 105, 107, 108, 130, 174–75, 204; vulnerability of, 175, 176, 195–96, 254n7, 259n71, 264n43
beach grasses, 5, 195; grazing on, 40, 45, 75, 76, 199
bedrock, 14–15, 16
Beebe family, 108, 120
beef cattle, 115, 199
Bell, Sheldon, 105
Berchen, William, 201
Beston, Henry, 10, 164
Biddle, Francis, 148
Biddle, Katherine Garrison Chapin, 148
Billingsgate Island, 4, 46, 66, 195
black flies, 22
black locust trees, 64, 171, 224n41, 254n67
Blake, Edwin, 140
boat- and shipbuilding, 42–44, 48, 56, 63–64, 77–78, 198, 233n130; forest products needed for, 43, 45, 55, 77, 79, 224nn41,43; decline of, 80, 88–89; switch of, to pleasure crafts, 89, 118–19, 199, 244n29
Boland, Edward, 164
bootlegging, 249n153
Boston, 14, 62, 77, 85, 119, 123, 150–51, 231n62; in colonial era, 33, 40; as market for Cape products, 40, 41, 42, 49, 53, 60, 80, 93, 115, 231n76, 236n16, 240n72; packet service to, 62, 72, 91; as fishing port, 85, 117; visitors to Cape from, 99, 104, 105, 122, 124, 128, 133–34, *135*, 145, 168, 226n70, 245n65, 246n97; retirees from, 130; commuters to, 151, 157
Boston and Sandwich Glass Company. *See* Sandwich Glass Works
Boston Fruit Company, 1
Bound Brook (Wellfleet), 64, 230nn47,59
Bourne, Richard, 34
Bourne, 34, 47, 151, 158, 161; summer estates in, 96, 108, 113, 123, 141, 246n79; water and sewage in, 180, 194. *See also* Bourne Bridge
Bourne Bridge, 159, 175
Bowditch, Ernest, 109
Bradford, William, 26, 32, 37
Brandeis, Louis, 139
Brewster, 68, 89, 120; natural setting of, 9–10, 16, 41, 125; growth of, 122, 157, 168. *See also* Nickerson State Park
brick making, 45
Brown, Theodore, 61
Buzzards Bay, 16, 17, 118, 126, 143; luxury development on, 98, 108–9, 120, 122, 124, 137, 138

Cahoon Barrel Factory (Harwich), 93
Camp Avalon, 133
Camp Cowasset, 133
Camp Edwards military base, 147, 194
campgrounds, 105, 203, 261n24. *See also* camp meetings
Camp-Meeting Grove Corporation, 100
camp meetings, 97, 100–102, 105
Cape Cod Advancement Plan, 144
Cape Cod Bay, 22, 43, 86, 118; development on, 124, 125, 158
Cape Cod Beacon, 164
Cape Cod Commission (CCC), 167, 168–69, 197, 209, 263n35
Cape Codder hotel (Falmouth), 134
Cape Cod Farm Bureau, 116, 143
Cape Cod Improvement Association, 127–28, 143, 197
Cape Cod Land Bank, 170
Cape Cod Mall, 155
Cape Cod National Seashore, 185, 203, 205–6, 210, 254n66; struggle for creation of, 163–66; land preservation by,

170, 179, 202, 204; and beach erosion, 196, 252n45; visitors to, 202, 262n32
Cape Cod Planning and Economic Development Commission (CCPEDC), 161–62, 167, 168–69. *See also* "Prospect: Cape Cod"
Cape Cod Poultry Organization, 116
Cape Cod Times, 182
Carbonell, Armando, 168
Carson, Rachel, 174
catboats, 58, 60, 63, 64, 89, 121; building of, 64, 89, 199; recreational, 89, 119, 136, 244n29
cattle, 115, 199, 243n19; in colonial era, 32, 33, 35, 38, 40, 46; overgrazing by, 35, 40, 46. *See also* beef cattle; dairy cows
cedar trees, 20, 25; uses for, 39, 63, 75, 89, 223n25
Centerville village (Barnstable), 63, 110, 126, 132, 155, 181
cereal crops, 49, 51, 226n70; short growing season for, 32; English settlers and, 32, 37, 38, 40, 223n29; mills to grind, 36, 38, 40, 69, 110; sale of, off-Cape, 68, 74, 231n76; decline of, 69, 74, 76, 81, 115, 199, 232n85. *See also* corn; wheat
Champlain, Samuel de, 24, 25, 26, 27
Channing, William Ellery, 8, 73
Chapman House, 115
charcoal, 44, 45
Chatham, 9, 120, 220n20; Native Americans in, 24; in colonial era, 39, 223n26; maritime industries in, 44, 63, 65, 67, 75–76, 83, 85–86, 236nn16,19, 259n7; transportation to and from, 91, 238n54; jobs in, 136; population growth in, 157, 158; water and sewage in, 180, 185, 192
—as tourist center, 126, 136; summer estates and hotels in, 98, 104, 106–7, 124, 130, 132, 139, 141; cottages in, 100, 125, 155
Chatham Associates, 124
Chatham Bars Inn, 124, 132
Chatham Country Club, 124, 132, 245n56. *See also* Eastward Ho Country Club
Chatham Hotel, 124, 245n55

Chatham Inn, 107
Chatham Monitor, 106, 107
Chequessett Inn (Wellfleet), 2–3, 97, 104, 134, 215n7; and Lorenzo Dow Baker's vision for Cape, 3, 97; amenities offered by, 104, 119, 124
chickens, 116, 186, 199, 208, 243n19; in colonial era, 32, 33; in nineteenth century, 67, 75, 90, 105, 198
Chocequoit Island, 123
Christian Churches of Southeastern Massachusetts, 101
Civilian Conservation Corps (CCC), 145, 164
Civil War, 82
clams, 41, 94, 104, 201, 208, 229n35; Native Americans and, 21, 23, 25, 27; as bait, 58–59, 76–77, 90
Cleveland, Grover, 96, 109, 122, 123
climate, 18, 20, 24, 32, 40, 81, 182. *See also* growing season; storms
clipper ships, 64, 78, 88, 231n62, 237n39
Coastal Wetlands Protective Act (Massachusetts, 1963), 177
Coastal Wetlands Restriction Act (Massachusetts, 1965), 177
Coastal Zone Management Act (CZMA, 1972), 189, 257n51
Coast Guard Beach, 196
cod fishing, 52–53, 58, 198, 227n10; on distant fishing grounds, 44, 48–49, 52, 55–56, 84, 87, 117, 235n14, 236n19; bait for, 49, 58–59, 63, 70, 117; by whaling vessels, 54, 228n24; long-lining in, 56–57, *58*, 235n14, 259n7; switch from, to mackerel, 84, 85, 86–87, 236n26; decline of, 84–85, 117, 118, 200
common law, 51, 174–75; and fish migration, 4, 70, 71, 232n92
Community Preservation Act (Massachusetts, 2008), 205, 253n63
commuters, 150, 151
Compact of Cape Cod Conservation Trusts, 170
Connors, Donald, 169
construction jobs, 147, 153, 157, 158, 162, 171, 261–62n25

continental drift, 13–15
Cook, Norman, 161
Coonamessett Pond, 70
Coonamessett Ranch (Falmouth), 147, 243n18, 249n154
Coonamessett River, 70
cordgrass (*Spartina alterniflora*), 19, 242n2
cordwood, 39, 69; growing scarcity of, 39, 47, 63, 65, 68, 78; industrial use of, 45, 65, 78; substitution of peat for, 45, 68, 231n80
corn, 49, 52, 76, 227n5, 231n76; Native Americans and, 13, 21, 22, 23, 24, 218n4; English settlers and, 32, 37, 38, 40, 49
Corn Hill Beach Colony (Truro), 105, 202, 203
Cory, Charles, 122
Cotuit village (Barnstable), 104, 126, 130, 132
crabs, 21. *See also* horseshoe crabs
Craigsville Beach, 101
Craigsville village (Barnstable), 101, 105, 125, 132
cranberries, 68, 92–93; market for, 68, 81, 92–93, 115, 139; wild, 68. *See also* cranberry bogs
cranberry bogs, 7, 68, 2, 139, 247n122; creation of, 68, 92, 232n83
Crane, Donald, 176, 177
Cronon, William, 216n16, 217n19
Crosby, Albert, 95, 104
Crosby, Jesse, 63, 77, 89
Crosby, Nathan, 95, 102, 104
Crowell, Levi, 62

dairy farms, 139, 154, 200
dams, 36; and fish migration, 36, 39, 69–70, 206, 232n93
Davis, Edmund, 108, 110
Davis, Wendell, 65, 227n1
DDT, 191, 257n57
deforestation, 2, 39, 45, 46–47, 48, 234n7; and soil erosion, 5, 45, 47, 73–74; extent of, by Thoreau's time, 5, 73–74; on Outer Cape, 47, 68, 73, 74; impact of, on Cape industries, 67, 77–78, 89; partial reversal of, 170–71

Dennis, 62–63, 158, 165, 176, 252n46; natural setting of, 9–10, 16, 47; farming in, 52, 139; forests and deforestation in, 73–74; summer visitors in, 104, 126, 137–38, 139; zoning in, 161, 185; land preservation in, 163–64, 252n46
—maritime industries in: salt making, 44, 49, 53, 64–65, 66, 89; fishing, 52, 62, 227n12, 231n64, 235n8, 236n26; boat- and shipbuilding, 64, 78, 88, 233n130
Dexter, Leonard, 103,
Dexter House Hotel (Falmouth), 103
diazinon, 191
Dickens, Monica, 191, 201
dikes, 71, 93–94, 174, 258n71; in Wellfleet, 4–5, 145–46, 165, 174, 178–79, 249n149; for railroad construction, 71, 93
diseases, 26–27, 219n4, 220n13
dories, 56–57, *58,* 60, 77, 85, 86, 235n14; building of, 43, 77
Dos Passos, John, 10, 148, 242n8
Dos Passos, Katharine, 10, 148, 242n8; Cape recollections by, 115, 126, 235n13, 244n29
Drake, Samuel, 119–20
Dukakis, Michael S., 169
dumps. *See* landfills
dunes, 164; formation of, 17; destabilization of, 17, 45, 46–47, 73, 74–75, 76, 176; efforts to protect, 46–47, 75, 76
Dwight, Timothy, 10, 39; observations on Cape by, 39, 47, 51–52, 53, 65, 73, 76, 202, 227nn2,10

Eastham, 58–59, 120, 195, 249n157; natural setting of, 9–10, 33, 166, 224n36; in colonial era, 27, 33, 38, 45, 46, 47, 224n37; forests and deforestation in, 33, 45, 46, 234n7; farming in, 47, 52, 68, 74, 115, 227n5, 242n5; camp meetings in, 97, 100; cottages in, 98, 137–38, 148; jobs in, 153, 158, 250n8; population growth in, 157, 158; land use issues in, 161, 185, 225n57, 226n67;

and Cape Cod National Seashore, 165, 166; water and sewage in, 187, 256n35
East Harbor (Truro), 75, 258–59n71
Eastward Ho Country Club (Chatham), 132, 240n75, 245n56. *See also* Chatham Country Club
Edwards, Agnes, 74, 129–30
eelgrass, 19, 93
eider ducks, 60, 77
Eldredge, Marcellus, 106–7
electricity, 142, 200
Ellis, Ephraim, 69
Ellsworth, George, 136
Environmental Defense Fund, 180
EPA. *See* U.S. Environmental Protection Agency
Erie Canal, 217n18, 232n85
eutrophication, 170, 180–81, 204

Fair Labor Standards Act, 140
Fall River, Mass., 105, 245n65
Falmouth, 132, 147, 158, 168, 161, 175; natural setting of, 9, 16; forests and deforestation in, 47, 94; maritime industries in, 57, 63, 65, 66, 77, 83, 89, 110, 115; farming in, 69, 115–16, 243n19; Milldam War in, 70, 207, 232n93; transportation to and from, 120, 159; jobs in, 137, 197, 205; as commercial center, 144, 148; water and sewage in, 146, 180, 190, 192, 194, 258n69; waste disposal in, 146, 186. *See also* Woods Hole
—summer accommodations in, 105, 120, 133, 141, 248n30; summer estates, 98, 108–10, 113, 120–21, 126, 130, 131–32, 139, 141; hotels, 103, 104, 108, 134
Falmouth Heights, 108
farming, 48, 116, 243n19; by Native Americans, 21, 22, 23, 24, 222n16; by English settlers, 32, 33, 38–40, 220n15; in early nineteenth century, 52, 67–69, 81–82; and off-Cape markets, 67–68, 115, 227n5, 231n76; challenges to sustainability of, 74–75, 79, 82, 98; diversification of, 81, 114–16, 208, 199–200; decline of, 99, 115, 136, 138–39, 154–55, 198–200, 251n17; and tourist economy, 114–16, 54–55, 138–39, 208; organic, 208. *See also* cereal crops; farmland; livestock
farmland: sale of, to developers, 108, 120, 154–55, 200; extent of, on Cape, 116, 155, 208
Faught, Millard, 241n106
Fay, Joseph, 108
Fay family, 120, 122. *See also* Fay, Joseph
fertilizer: fish and crabs as, 23, 24–25; animal manure as, 33, 38, 48, 49, 115, 233n103; artificial, 94–95, 109, 120, 182, 191
Finch, Robert, 10, 263n41
fires, 166, 171, 216n14, 221n1; Native Americans' use of, 22, 216n14, 219n5, 221nn1,8
firewood. *See* cordwood
fishing, 44, 53, 71–72, 80, 198, 226n64, 263n38; for anadromous fish, 4, 5, 48, 59, 84, 216n10; bait used in, 4, 49, 58–59, 63, 70, 76–77, 228n31; by Native Americans, 22–23, 25; vessels used in, 43, 48, 53, 56–57, 117, 200–201, 235n14 (*see also* schooners); on distant grounds, 60, 117 (*see also* fishing banks); decline of, 75–76, 84–87, 117–18, 200, 235n14, 236n19, 239n65; recreational: *see* sport fishing. *See also* cod fishing; long-lining; mackerel fishing pound nets; seines; weirs
fishing banks, 14, 48, 55–56, 75–76, 81, 228n27; origins of, 17–18; kinds of fish sought on, 44, 48, 55, 56, 84; dangers of, 55–56, 86. *See also* Georges Bank; Grand Banks
fish processing, 84, 120, 121, 174; as deterrent to tourism, 120, 174
"flakes" (fish drying racks), 2, 45, 57, 84, 224n51, 225n58
flounder, 85, 118, 136n22
forests, 171; before white settlement, 5, 13, 31, 33, 216n14; composition of, 5, 20, 22, 25, 31, 44, 171, 216n14, 221–2n8, 254n67; fires and, 22, 25, 171, 216n14, 219n5, 221n1, 221–22n8, 254n66; efforts

forests (*continued*)
    to protect, 39, 45, 46–47, 48; multiple uses made of, 43–44, 45, 48, 75, 94 (*see also* cordwood; lumber); on Outer Cape, 45, 47, 68, 73–74, 166; partial recovery of, 170–71. *See also* deforestation
Frazier, Charles, 165
Freeman, Frederick, 69–70, 74, 78

gasoline, 180
glaciers, 14–15; shaping of landscape by, 9–10, 13, 14–17, 25, 37
glass manufacturing, 47, 94, 238n58
Gloucester, Mass., 56, 85, 117, 243n24
golf, 132, 155, 156; miniature, 155, 156–57, 248n132; environmental costs of, 182, 183, 191
Gosnold, Bartholomew, 26
grain. *See* cereal crops
Grand Banks, 14, 44, 51, 55, 62, 117, 228n27
Great Depression, 134, 145, 147
Great Island (Wellfleet), 75, 166, 178–79
Great Meadow (Barnstable), 71, 82, 234n5
Great Salt Marsh (Barnstable). *See* Great Meadow
groins, 196
groundfish, 118
growing season, 23, 24; for grains, 32
guesthouses, 103, 124
Gulf Stream, 14, 17–18, 175

haddock, 55, 56, 58, 84
Hamilton, Alice, 133
Hammatt, Mary, 133
Hannah, Samuel, 140, 153–54
harbors, 60–63, 230n47, 249n153; and sand erosion, 46, 75–76, 86, 230n59, 236n26, 237n48, 237–38n49; shallowness of, most, on Cape, 54, 63, 91, 117; centrality of, to nineteenth-century Cape, 60–63, 72; pollution in, 182. *See also specific harbors*
Hardy, Charles, 124, 141
Harvard University School of Public Health, 180, 185

Harwich, 120, 158, 185, 215n1, 229n38, 258n63; in colonial era, 46, 54; maritime industries in, 54, 63, 65, 77, 93, 236n26; summer accommodations in, 98, 122–23, 126–27, 132, 137, 140, 148
Hatches Harbor (Provincetown), 179
Hay, John, 10
heath hens, 22, 25, 222n15
Herring River (Wellfleet), 4, 179, 208; and shellfish, 4, 9, 208; fishing on, 4, 48, 59, 84, 216n10; controversial dike on, 4–5, 145–46, 165, 174, 178–79, 249n149
"herring runs," 4, 41, 84
Higgins, Simeon, 91
Highlands House (Truro), 125, 129
highways, 159, 193, 246n77; and transformation of Cape, 114, 159; congestion on, 147, 159, 168, 251–52n30. *See also* Mid-Cape Highway; Route 6A; Route 28
historic districts, 162–63
Hodgson Homes, 140
home prices, 203, 204, 261nn18,24
Homestead Trust, 140, 153–54
Hopkins family, 134–36
horses: in colonial era, 32, 33, 35, 40, 225n58; in nineteenth century, 75, 105, 126, 222n19, 243n14; in twentieth century, 136, 243n19, 259n4, 260n12
horseshoe crabs, 22, 24–25, 32–33, 233n103
Human Society, 47
Hunt, Thomas, 27
hunting, 21, 22, 240n72. *See also* sport hunting
hurricanes, 6, 175–76, 254nn5,6, 264n43; in colonial era, 32, 35, 221n3; in nineteenth century, 66, 230n5
Hyannis, 122, 197, 205; Native Americans in, 25; transportation to and from, 90–91, 104, 133; tourism in, 105, 123, 148, 261n23; as commercial center, 144, 148, 154, 155, 158, 161, 168. *See also* Hyannis Port
Hyannis Port, 107, 110, 125, 126, 130, 131–32, 164
Hyannis Port Land Company, 107

INDEX

271

immigrant workers, 197
insects, 19, 3, 22, 145, 191; use of pesticides against, 191, 257n57. *See also* mosquitoes
ironworks, 45
Iyanough House (Hyannis), 104

Janes, E. C., 129, 134
Jefferson, Joseph, 109
Jones, Charles, 108, 109
Jones Act (1962), 177
Jordan, Eben, 106–7

Keith, Hastings, 164
Keith Car and Manufacturing Company, 92
Keith family, 108. *See also* Keith, Hastings
Kelley, Stillman, 62–63
Kendall, Edward Augustus, 73, 74, 233n103
Kennedy, John F., 164, 165

Labrador Current, 17, 18
land prices, 109, 137, 141, 144; in decades since World War II, 154, 165, 167, 172, 201, 202, 203. *See also* land speculation
land values. *See* land prices
landfills, 146, 186–87, 191, 257n46; strained capacity of, 146, 186–87; leaching from, 187, 256n35
livestock, 32, 33, 35, 40, 73, 199, 225n58. *See also* cattle; chickens; horses; pigs; sheep
lobsters, 23, 41, 49, 118, 130
Long Island, 15–16, 119, 126, 209, 264n43
long-lining, 56–57, 58, 235n14, 259n7
"Lower Cape" (term), 1, 95, 215n1
lumber, 43–44, 63, 74; cost of, 39, 47, 66, 77, 80, 88; for homes, 39, 223n23; for boat- and shipbuilding, 43, 55, 64, 77, 80, 88; growing shortage of, 43, 64, 74, 77, 78–79, 89; for saltworks, 66, 78, 89, 226n69, 237n42
Luscombe, Earl, 165
Lyman, Theodore, 99, 240n73

MacKaye, Benton, 144

mackerel fishing, 49, 53, 58, 59–60, 85–87, 200, 236nn26,28; vessels used in, 60, 63, 76, 85, 86; differences between, and cod fishing, 60, 85, 86; bait used in, 70; decline of, 84, 86–87, 117, 200, 239n65, 259n6; new methods of, 85–86, 230n55, 236n22; efforts to regulate, 226n64
Magnuson-Stevens Fishery Conservation and Management Act of 1976, 200
Mailer, Norman, 10
maize. *See* corn
Marblehead, Mass., 119
Marshall, Anthony, 241–42n2, 243n20
marshes, 3, 19, 23, 174; formation of, 19; ecological importance of, 20, 93, 238n55; conversion of some, to cranberry bogs, 68, 92, 232n83; draining of, 71, 145–46, 174, 178, 254n1 (*see also* dikes); efforts to protect, 163, 177–78. *See also* salt hay
marsh grasses, 3, 19, 25. *See also* salt hay
Marston Mills village (Barnstable), 64, 70
Martha's Vineyard, 15, 16, 119, 169, 204; ferry service to, 62, 91
Mashpee, 157–58, 168, 180, 235n12, 243n18; Native Americans in, 27, 202, 225n62
Mashpee River, 181
Massachusetts Audubon Society, 163, 170, 202, 206
Massachusetts Bay Colony, 33, 40, 46, 206
Massachusetts Bureau of Statistics of Labor, 94, 105
Massachusetts Department of Commerce, 161
Massachusetts Department of Environmental Quality Engineering (DEQE), 192, 194
Massachusetts Department of Labor and Industries, 105, 199
Massachusetts Department of Natural Resources, 179
Massachusetts Department of Public Works (DPW), 177, 179, 193–94
Massachusetts Highway Commission, 129, 246n77

Massachusetts Historical Society, 49, 52, 53, 54, 61–62, 65
Massachusetts Merchant Marine Academy, 143
Massachusetts Water Resources Commission, 176–77
Massasoit, 219n4
McCann, James, 180
McCarthy, Mary, 10, 148
McKay, William, 84
McKenzie, Matthew, 236n24
Megansett Beach (Falmouth), 108
Megansett Shores (Falmouth), 108
Mellen, Reverend, 61–62
Melville, Herman, 37, 228n25
menhaden, 22–23, 55, 228–29n32; fishing for, 23, 58, 85, 86; as fertilizer, 23, 94–95
Merchant, Carolyn, 217n19
Methodist camp meetings, 97, 100–102
Mid-Cape Highway, 16, 155, 159, 162, 168, 193
migratory fish. *See* anadromous fish
milk cows, 75, 82, 199, 259n4. *See also* dairy farms
Milldam War (Falmouth), 70, 207, 232n93
Millenium Grove, 100–101
mills, 36, 38, 40, 64, 222n14; and migratory fish, 36, 39, 48, 69–70, 207, 232n92; for cloth making, 59, 69–70, 207, 232nn86,89. *See also* windmills
Monomoy Point, 195, 237n48
Monument Neck (Bourne), 109, 123
mosquitoes, 3–4, 126, 215n7; as deterrent to tourism, 3, 126, 145–46; efforts to control, 4, 145–46, 249n149, 254n1
motels, 183, 251n21, 261n23; spread of, from late 1930s on, 138, 139–40, 154, 155–56, 261n23; newer and fancier, 197, 203
Murdoch, Alice, 133

Nantucket, 15, 55, 204; ferry service to, 62, 91, 104; whaling from, 83
Nantucket Shoals, 15, 53, 56, 85
Nantucket Sound, 22, 118, 125, 263n35; formation of, 16, 17; fishing in, 63, 86; development on, 107, 124, 131–32, 137, 138
National Geographic, 168
Native Americans, 6, 24–27, 32, 85, 219n8; food sources utilized by, 21, 22–23, 41, 59, 68, 85, 219n8, 220n5; use of fires by, 22, 216n14, 219n5, 221nn1,8; and diseases, 26–27, 219n4, 220n13; in Mashpee, 27, 202, 225n62; and English settlers, 32, 36–37, 41, 219n4, 222n16, 225n62
Nauset area, 21, 37–38, 195; Native Americans in, 21, 24
New Bedford, Mass., 55, 62, 83, 235n10
Newport, R.I., 102, 119, 126, 204
New York City, 123; as market for Cape products, 41, 49, 80, 93, 236n16, 240n72; visitors to Cape from, 99, 104, 122, 124, 133, 245n65, 246n97; retirees from, 130
Nicholson, Donald, 164
Nickerson, James, 128, 162, 197
Nickerson, James, Jr., 187–88, 252n36
Nickerson, Joshua, 138, 153
Nickerson, Joshua, II, 141, 165–66, 245n55
Nickerson, Roland, 122
Nickerson Neck, 106
Nickerson State Park (Brewster), 145, 163, 202, 203, 206
Nobscussett House (Dennis), 104, 134, 139

oak trees, 20, 39, 44, 171, 216n14, 221n8; uses for, 43, 44, 63, 77, 89; scrub, 171, 254n67
Old Colony railroad, 93, 104, 105, 110, 120, 126; shareholding in, 237n45, 238n54
O'Neill, Thomas "Tip," 164
orchards, 40, 116, 198
organic farms, 208
Orleans, 120, 157; natural setting of, 20, 164; maritime industries in, 65, 66, 76–77, 87, 235n9; transportation to and from, 91, 145, 159, 162; development in, 137, 138, 141, 148, 158, 168, 203; farming in, 231n76

INDEX

Orleans Associates, 138, 141
Osterville village (Barnstable), 63, 104, 110, 126, 130, 132
Otis Air Force Base, 180, 194–95
Outer Cape, 18, 20, 69, 91, 162; Thoreau on, 5, 73; changing shore line of, 8, 196, 217n17, 258n70, 264n43; formation of, 16, 17; environmental damage on, 45, 47, 68, 74–75, 231n77, 234n7; growing population of, 158, 167; preservation efforts on, 170, 179, 196 (*see also* Cape Cod National Seashore); water and sewage on, 182, 184, 185, 187, 192, 195, 256nn35,39, 261n21. *See also* Eastham; Provincetown; Truro; Wellfleet
overgrazing, 75, 76; in colonial era, 35, 40, 46–47, 225n57, 263n34; on dunes, 45, 46–47, 73, 74, 207
oysters, 41, 118, 130, 208; early abundance of, 23; farming of, 36, 60, 229n43, 263n36; sale of, off-Cape, 41, 53, 60, 93, 120; overharvesting of, 41, 118; recovery of, 60, 118, 178, 224n27, 256n29; and water quality, 182; regulation of, 226n63, 229n43, 244n27, 256n29

packet services, 49, 61–62, 72, 226n70; decline of, 90, 91, 237–38n49
Palfrey, John Gorham, 53, 222n16
Pamet Harbor (Truro), 75
parking lots, 159–60, 168, 196
peat: formation of, 17, 20, 68, 231n79; use of, for burning, 45, 68, 231n80
Pendergast, James, 261n23
Penzance Point, 109
Perry, Ezra George, 96, 108, 127, 128; as developer, 95, 96, 109, 121, 123, 239nn60–62, 245n54; vision of, for Cape, 96, 97–98, 109–10, 113, 127, 187; background of, 96, 109, 239n60; promotional writing by, 96, 115, 118, 119, 121, 123, 239n62, 245n54
pesticides, 182, 191, 257n17
Phinney Harbor (Bourne), 96
phragmites, 93–94, 238n56

Piercy, Marge, 10
pigs, 67, 75, 186, 198; in colonial era, 32, 33, 34–35, 59
Pilgrims, 26–27, 31–32, 33–38, 41, 59, 220–21n15; landscape found by, 5, 31, 33, 73, 74; and Native Americans, 26–27, 32, 36–38, 219n4, 220nn6,14
pine trees, 39, 94, 221n8, 233n106; white, 19–20, 43, 44, 171, 223n25, 233n106; use of, in maritime industries, 43, 45, 63, 65, 66, 67, 77, 89, 224n43, 237n43. *See also* pitch pines
pitch pines, 22, 44, 47, 171, 216n14, 233n106; and fires, 22, 171, 216n14, 221n8; limited usefulness of, 43; prevalence of, on twentieth-century Cape, 171, 254n67
Pleasant Bay, 106, 119, 133, 158
Pleistocene epoch, 14–15
Plymouth Harbor, 31, 37
ponds, 16, 17, 99, 125, 180; and fish migration, 5, 22, 59, 69, 70, 178, 179, 216n10, 219nn7,8; formation of, 16; ice from, 81, 84; degraded water quality in, 146, 180–81; sluiceways between, 219n8, 229n40
population, 117, 145; growth of, in early eighteenth century, 80–81; decline of, in post–Civil War decades, 81, 87, 121, 145; renewed growth of, after 1920, 113–14, 121, 138, 145, 147; summertime swelling of, 137, 147, 158; in post–World War II decades, 152, 157–58, 161, 205, 250n7, 260n14, 262n27; changing composition of, 153, 158, 160, 251n24, 259–60n12, 260n14; density of, 252n34
Portanimicut Camp, 133
Portnoy, John, 179
Portuguese, 115, 117, 148, 242n11
pound nets, 58, *61*, 85–86, 117, 120; types of fish caught in, 58, 85–86, 136, 236n22
Prence, Thomas, 33
"Prospect: Cape Cod," 167, 168–69, 197, 253n51
Providence, R.I., 62

Provincetown, 105, 120, 126, 161, 206, 262n30; natural setting of, 9, 41, 47, 195; shifting sands in, 46, 76, 195, 196, 230n59, 258–59n71; transportation to and from, 90, 91, *135,* 159; population trends in, 117, 157; cottages in, 148; water and sewage in, 184, 185, 192, 195, 256n39, 261n21; in colonial era, 206, 224n50, 225n57. *See also* Provincetown Harbor
—maritime industries in: fishing, 8, 44, 48–49, 52, 53, 72, 84, 117, 148, 157, 200, 227n10, 236nn16,19, 263n38; boat- and shipbuilding, 43, 56, 63, 77, 88, 89; whaling, 54, 83, 235n12; salt making, 57, 65, 66, 78, 89; fish processing, 84, 121
Provincetown Art Association, 148
Provincetown Harbor, 46, 76, 91, 206; and sand erosion, 46, 76
Public Works Administration (PWA), 138, 145
purse seines, 86–87, 230n55
Putnam's Magazine, 8

Quanset Sailing Camp, 133

railroads, 90–93, 123–24, 237n46, 245n65; economic impact of, 86–87, 90; extension of, across Cape, 90–91, 99, 100; replacement of packet services by, 91, 237–38n49; and expanded market for Cape products, 92–93, 120, 236n16; environmental impact of, 93–94, 174, 238n55; financing of, 93, 237n45, 238n54; and increase of visitors, 99, 100, 103–4, 105, 126, 133–34; decline of, on Cape, 134, 159
recreation, 107, 126–27, 133, 155, 173. *See also* golf; sport fishing; sport hunting; sport sailing
Redding, George, 147
resin, 45, 224n43
Resource Conservation and Recovery Act (1975), 187
restaurants, 142, 144, 158, 203, 208; seafood in, 118, 130, 136, 208; jobs in, 137, 201; fake sea themes for, 149, 157, 160
retirees, 148, 153, 162, 204; growing numbers of, on Cape, 138, 148, 152, 153, 157, 158, 171; and housing markets, 138, 151, 155, 250n5
river herring, 4, 59, 70, 84, 228–29n32, 229n40; as bait, 58, 70, 228n31
roadside stands, 116, 243n15
rope making, 64, 77, 91, 198
Route 6, 159, 178. *See also* Mid-Cape Highway
Route 6A, 159, 162–63, 193
Route 28, 16, 193; development along, 138, 139, 155–56, 163; traffic on, 147, 159
Roy, Steve, 193–94
Russo, Richard, 10
Ryder, Marion Crowell, 62
rye, 49, 52, 69, 198; English settlers and, 32, 37, 38, 40

Sagamore Bridge, 159, 168, 175, 251–52n30
sail making, 64, 198
salt hay, *50,* 71, 75, 222n20, 227n5; English settlers and, 33, 35, 38; sale of, off-Cape, 227n1
salt making, 44, 49, 64–67, 174, 226n69, 231n66; need for, to preserve fish, 44, 64, 231n64; wood consumed in, 44, 67, 78, 79, 226n69; tariff protection for, 49, 65, 78, 80; as dominant feature of shoreline, 49, *67,* 121; economic importance of, 53; decline of, on Cape, 78, 80, 89–90, 121, 237nn42,43
salt meadow grass (*Spartina patens*), 19. *See also* salt hay
Saltonstall, Leverett, 164, 165
saltwater infusion, 182–83, 184, 185, 193–94, 256n39
sand erosion. *See* soil and sand erosion; beach erosion
Sandwich, 37, 47, 140, 163; Native Americans in, 25; in colonial era, 34–36, 48, 224n37, 225n57, 226n70; farming in, 35, 36, 51; mills in, 36, 64, 69; fish migration in, 36, 69–70; transportation to and from, 36, 90, 99,

159, 226n70; maritime industries in, 43, 63, 65, 77, 89; glass manufacturing in, 47, 94, 238n58; in decades since World War II, 151, 158, 160, 162, 180, 192, 194
Sandwich Glass Works, 47, 94, 238n58
Santuit House (Cotuit), 104
Sargent, Francis, 164
schooners, 1, 2, 43, 44, 57, 61, 72; building of, 43, 63–64, 77, 88–89, 237n36; changes in, 56–57, 235n14; as packet ships, 62, 91, 226n70; eclipse of, 76, 77–78, 91, 117, 187; and mackerel, 85, 236n28
sea levels, 14, 16–17, 195, 218n2, 258n70
Sears, John, 65, 74, 226n69
Sears, Nathan, 63
Sears, Seth, 62–63,
seines, 49, 58, 226n64. *See also* purse seines
septic systems, 140, 150, 170, 185, 203, 261n21; pollution from, 180–86, 192, 194; efforts to regulate, 180, 182, 185, 202
service jobs, 7, 136–37, 201
Sesuit Club, 132
sewer systems, 146, 183, 192
Shawme-Crowell State Forest, 145, 163, 202
Shay, Edith, 115, 126, 235n13, 244n29
sheep, 69, 199, 207, 232n89; in colonial era, 33, 35, 40, 45; overgrazing by, 40, 45, 74–75, 199, 207; decline of, on Cape, 115, 199
shellfish harvesting, 47–48, 94, 204, 207–8; by Native Americans, 21, 22, 23, 24, 38, 218n4; and aquaculture, 60, 208, 229n43, 263n36. *See also* clams; crabs; lobsters; oysters
shipbuilding. *See* boat- and shipbuilding
Shiverick ship works (Dennis), 64, 88
shopping centers, 155
Silver Beach (Falmouth), 108
skiffs, 55, 89, 97; recreational, 119, 141, 142, 199
sloops, 54, 55
Smith, John, 41

soil and sand erosion, 74–76, 132; deforestation and, 5, 45, 47, 73–74; efforts to reduce, 46–47, 76; impact of, on harbors, 46, 75–76, 86, 230n49, 236n26, 237n48. *See also* overgrazing
Sparrow, Rich, 71
sport fishing, 99, 123, 136, 153
sport hunting, 99–100, 107, 122
sport sailing, 118–19, 121. *See also* yacht clubs; yachts
Squanto, 32
Stage Harbor (Chatham), 24
Standish, Myles, 26
steamboats, 91
storms, 175, 229–30n45, 264n43; northeastern, 18, 32, 66, 78, 175–76, 252n45, 255n15, 264n43; on fishing grounds, 55, 56; increased vulnerability to, 175–76, 177, 195, 196, 209. *See also* hurricanes
stoves, 69, 142
Strait of Belle Isle, 52, 53, 54, 60, 85
Strong Island (Chatham), 106–7
summer accommodations. *See* campgrounds; camp meetings; motels; summer camps; summer cottages; summer estates; summer hotels
summer camps, 133, 158
summer cottages, 97, 98, *106*, 125; on camp meeting sites, 97, 100, 101, 102; Lorenzo Dow Baker and, 97, 102, 104, 125, 202; on hotel or inn grounds, 104, 134; colonies of, 105, 202; growth of, in decades after 1920, 113, 129–30, 134, 137–38, 139–42, 143, 148; upper-end, 124, 147; since World War II, 153–56, 158, 185, 203, 261n24; upgrading of, 203, 261n24; prefabricated, 247n124; rental prices for, 248n130, 261n24
summer estates, 108–10, 134, 141, 147–48, 202, 241n106; service jobs on, 7, 136–37, 201; promotion of, 96, 98, 109–10, 119–24; growth of, on Buzzards Bay and south shore, 108–10, 130–32; tax revenue from, 129, 130, 139, 246n79; travel to, 245n65

summer hotels, 95, 103–4, 115, 124–26, 129; emergence of, 98, 103–4, 105; wealthy clientele of, 103–4; "American plan" for, 104, 113, 125, 134, 142; limited success of, 107, 108, 125–26; decline of, 134, 142. *See also* Chequessett Inn
Swift, Gustavus, 102
swordfish, 118, 136

tariffs, 49, 65, 78, 80
Tarr, Joel, 186
Tawasentha, 95, 104

Thompson, Elroy, 126
Thoreau, Henry David, 3, 8, 98, 103, 128, 216n12; times spent by, on Cape, 5, 8, 125; characterizations of Cape by, 5, 98, 100, 175; on abundance of sea life, 23, 58, 229n35; on human impact, 26, 70, 73–74, 77–78, 210; on Cape's future, 98, 110, 195, 203; on Cape's natural beauty, 101, 110, 196, 201, 259n9
Thoreau, Sophia, 8
tidal flooding, 175–76
tourist economy, 6, 7, 113–14, 119–28, 130–42, 201–4, 209; early visions of, 3, 5, 96–98, 109–10, 113, 127, 187; and job market, 7–8, 136–37, 153, 158, 162, 171, 197, 201, 205; and transformation of shoreline, 61, 98, 107, 110, 174–77; mix of classes in, 97–98, 113–14, 137–38, 151, 202; farmers and, 114–16, 54–55, 138–39, 208; and maritime industries, 117–19, 130, 199, 208, 244n29; deterrents to, 120–21, 125–26, 145; and land prices, 128–29, 138–39, 145–46, 172, 196–97; automobiles and, 128–30, 147, 159–60; worries generated by, 143–44, 160–64, 168–69, 172, 201, 209; and postwar prosperity, 147–49, 151–60; and Cape's environment, 160–72, 174–97. *See also* recreation; restaurants; summer accommodations
traffic, 147, 159, 162, 168, 251–52n30
transportation. *See* automobiles; highways; packet services; railroads
trees. *See* cordwood; deforestation; forests; *specific types of trees*
Truro, 87, 120, 157; in colonial era, 27, 39, 40, 42, 47, 223n26; forests and deforestation in, 39, 47, 73, 74, 171; maritime industries in, 44, 56, 63, 65, 66, 77, 86, 87, 121, 236n16; farming in, 69, 76, 105, 115, 116, 241–42n2; sand erosion in, 75, 195; summer accommodations in, 98, 103, 105, 121, 125, 129, 137–38, 140, 164, 202; transportation to and from, 159, 178, 237n49; and Cape Cod National Seashore, 165, 166; land use issues in, 170, 179, 260n16, 261n20; water and sewage in, 185, 192, 256n39
Tsongas, Paul, 168
tuna, 85–86, 136

U.S. Commission of Fish and Fisheries, 131
U.S. Environmental Protection Agency (EPA), 182, 189, 194, 205
U.S. National Park Service, 164–65, 166, 179, 206, 262n33. *See also* Cape Cod National Seashore
Underwood, Emily Watson, 100
Underwood, Loring, 100, 113, 240n75
United Fruit Company, 1

Vorse, Mary Heaton, 10, 148

Wampanoag Indians, 24, 219n4
waste disposal, 186, 256n40, 257n47. *See also* landfills; septic systems; sewer systems
Water Quality Management Plan, 182, 183, 185, 190, 191, 192; lot-size recommendation of, 180, 184, 185
water supply, 31, 167, 182–86; threats to, 163, 185–86, 192–93, 195, 256nn34,35,42 (*see also* saltwater infusion); efforts to protect, 163, 185, 252n41, 258n61; growing demands on, 183. *See also* aquifer
water temperatures, 124

INDEX

277

Webster, Daniel, 52, 99, 240n73
weirs, 244n41; Native Americans and, 22–23, 25; offshore, 230n55. *See also* pound nets
Wellfleet, 72, 122, 259n4, 260n12; natural setting of, 3, 9, 224n36; Native Americans in, 24, 38; in colonial era, 37–38, 225n57, 226n64; forests and deforestation in, 47, 73, 75, 78; harbor silting in, 75, 230n47; population decline and rise in, 81, 87, 90, 148, 157, 158; transportation to and from, 91, 93, 129, 248n137; summer accommodations in, 97, 110, 120, 122, 133, 137 (*see also* Chequessett Inn); and Cape Cod National Seashore, 165, 166, 168. *See also* Herring River
—maritime industries in: oyster harvesting, 41, 60, 118, 178, 182, 226n63, 229n43, 244n27, 256n29; whaling, 42, 52, 54, 55, 83, 228n24; fishing, 44, 52–53, 84, 85, 86, 87, 228nn24,31, 236nn21,28, 239n65; wharves, 61, 230n47; boat- and ship-building, 64, 88, 119, 228n26; salt making, 65, 66, 89
—emerging issues in: clean water supply, 182, 184, 187; waste disposal, 187, 257n46; building regulations, 190, 257n54, 260–61n16, 261n20; home prices, 203, 260n12, 261n18
wells, 190, 192–93, 256n39; private, 146, 184, 185, 187, 256n38
West Indies, 44, 53, 56, 82, 85
Wetlands Protection Act (Massachusetts, 1972), 177, 260n13
whale oil, 2, 41–42, 54–55, 83, 228n22, 235nn9,10
whales, 41; beached, 38, 41, 42, 224n37. *See also* whale oil; whaling

whaling, 2, 41–42, 48, 54–55, 72, 82–83; vessels used in, 52, 54–55, 77, 83, 228n26; decline of, 54–55, 76, 82–83, 110, 117, 228n24, 235nn9,10,12; multinational crews in, 117, 197, 202
wharves, 44, 50, 60–61, 174–75; collapse of many, 2, 80, 87, 121
white pine, 19–20, 43, 44, 171, 223n25, 233n106
Whitlock, Herbert, 252n36
Whitman, Levi, 52, 54
Wianno, 126, 132
Wilson, Edmund, 10, 148
windmills, 40, 65, 68, 75, 89, 110, 226n69; replicas of, 157, 160, 248n132
wind power, 263n35. *See also* windmills
Winsor, Christopher, 142, 248n137
Wirth, Conrad, 165
wolves, 34, 35, 223n21
woods. *See* forests
Woods Hole, 94–95, 131, 132, 148, 192, 246n92; summer estates in, 108, 120
World War II, 147, 149, 150
Wychmere Harbor, 126–27
Wychmere Pond (Harwich), 126–27

yacht clubs, 96, 97, 114, 132
yachts, 118–19, 132, 136. *See also* yacht clubs
Yarmouth, 62, 158, 165, 192, 223n28; natural setting of, 9–10; Native Americans in, 25; maritime industries in, 44, 53, 61, 64, 65, 66, 85, 89–90; environmental damage in, 46, 73–74, 190, 192; visitors in, 62, 97, 98, 100–101, 137, 155–56; transportation to and from, 159; land use issues in, 161, 162

zoning, 155, 161, 162, 163, 190, 203, 251n21; opposition to, 161, 162, 169, 190

JOHN T. CUMBLER is a professor of environmental and social history at the University of Louisville, where he has won several distinguished teaching awards. He was the John Adams Distinguished Fullbright Fellow at the University of Groningen, the Netherlands, and is an Honorary Fellow at the University of Warwick, UK. Cumbler has published several books on American social, economic, and environmental history, including *Reasonable Use: The People, the Environment, and the State, New England 1790–1930* and *From Abolition to Rights for All: The Making of a Reform Community in the Nineteenth Century*. He earned a PhD at the University of Michigan under Sam Bass Warner. He lives eight months of the year on Cape Cod, where he builds boats and rescues large marine mammals and sea turtles.